THE DRIVING F

EVOLUT

Genetic Processes in Populations

THE DRIVING FORCES OF EVOLUTION

Genetic Processes in Populations

David Wool

Professor (Emeritus) of Zoology
Tel Aviv University, Israel

CRC Press
Taylor & Francis Group
Boca Raton London New York

CRC Press is an imprint of the
Taylor & Francis Group, an **informa** business
A SCIENCE PUBLISHERS BOOK

First published in 2006 by Science Publishers

Published 2019 by CRC Press
Taylor & Francis Group
6000 Broken Sound Parkway NW, Suite 300 Boca
Raton, FL 33487-2742

© 2006, copyright reserved
CRC Press is an imprint of Taylor & Francis Group, an Informa business

First issued in paperback 2019

No claim to original U.S. Government works

ISBN-13: 978-0-367-45392-3 (pbk)
ISBN-13: 978-1-57808-445-6 (hbk)

**Visit the Taylor & Francis Web site at
http://www.taylorandfrancis.com**

**and the CRC Press Web site at
http://www.crcpress.com**

Library o o o ongross o ataloging-in-Publication Data

Wool, Daddd.

Tde drdddng forces of edolutdon: genetdc processes dn populatdons/Daddd Wool.

    p.cm.

Includes bdbldograpddcal references (p.).

ISdN 1-57808-445-8

1. Populatdon genetdcs. 2. Edolutdonary genetdcs. I. Tdtle.

QH455.W66 2006

576.5'8--dc22

2006048493

Preface

The material for the present book was first prepared over the course of 12 years of teaching of three courses – Population Genetics, Ecological Genetics, and Evolution – at the Faculty of Life Sciences, Tel Aviv University (the Ecological Genetics course continued for 15 more years). This was a period of deep controversy in the field of evolutionary theory over the importance of random processes, following the introduction of the Neutral Theory by Motoo Kimura in 1968. This controversy generated hundreds of publications on different aspects of evolution, and my lectures had to be frequently updated to give the students a concise and objective view of the current status of the theory.

The book was originally published (in Hebrew) in 1985. It is reassuring to see that the theoretical framework of the book, which represents the core of the quantitative theory of evolution, is as relevant today as it was 20 years ago (what has changed dramatically in these years is the volume of data in support of the theory. I updated the selected examples in 1998, and again in 2005). In fact, the reference books which I had used when I myself was a student of these subjects – such as Li (1956), Crow and Kimura (1970), and Wallace (1969) – are as applicable as ever for the purpose, although they may be difficult to get hold of today. This is proof of the essential stability of the theoretical basis of evolution. A recent theoretical book, which is often consulted in American universities, is that by Hartl & Clark (3rd edition, 1997). Readers who are interested in a wider scope of evolution, including aspects of palaeontology, taxonomy and phylogeny, may consult Price (1996).

I am grateful to Varda Wexler, who spent many hours improving and redrawing the illustrations, and to Naomi Paz for the meticulous editing of the text. I hope that the many colleagues whose publications I have cited, will find no reason to object to my interpretation of their work.

Ramat Hasharon, Israel 2005   **David Wool**

LITERATURE CITED

Crow, J.F. and M. Kimura. 1970. An Introduction to Population Genetic Theory. Harper & Row, New York, USA.

Graur, D. and W-H. Li. 2000. Fundamentals of Molecular Evolution. Sinauer, Sunderland, USA.

Hartl, D.L. and A.G. Clark. 1997. Principles of Population Genetics. 3rd edition. Sinauer, Sunderland, USA.

Li, C.C. 1956. Population Genetics. University of Chicago Press, Chicago, USA.

Price, P.W. 1996. Biological Evolution. Harcourt-Brace-Saunders Publishers, Orlando, USA.

Wallace, B. 1969. Topics in Population Genetics. Norton, New York, USA.

Contents

Part II: Selection in Nature

Part III: Macro-evolution

Introduction

Almost 150 years passed since the publication of Darwin's book, "The Origin of Species", in 1859. Intensive research in different fields of biology during this time interval has resulted in a deeper and wider understanding of comparative anatomy, physiology, and behavior of animals and the accumulation of data on fossil organisms (palaeontology). A revolution in the knowledge and understanding of the processes of heredity began with the re-discovery, in the year 1900, of Gregor Mendel's work on hybridization of peas, and the birth of the science of genetics. Quantitative theories of population genetics and evolution, built on the basis of Mendelian genetics, began to emerge in the 1920s.

These theories – supported by experimental laboratory studies on mutations, particularly of fruit flies *(Drosophila)*, by T.H. Morgan and his associates, gave a fresh start to the theory of evolution, which had had the character of "armchair biology" in the 19th century.

With the discovery of the structure of the genetic material, DNA, in the 1950s, and the understanding of the physical basis of mutations, the theory of evolution again had to adjust. The development of molecular evolution as a sub-discipline led to the re-examination of current phylogenetic assumptions concerning different animal groups.

Many books on evolution were published in the last five decades. Some of these books are strictly theoretical, others are descriptive and limited to particular biological phenomena. Some are designed as more-or-less comprehensive textbooks. I have called the present book "The Driving Forces of Evolution" to emphasize that the historical aspect of evolution – the description of the lines of descent connecting taxa – (phylogeny), which occupies the central part of many texts on evolution – is only briefly discussed in the book. Rather, I deal almost exclusively with the forces which drive the evolutionary process, cause genetic changes in populations, and regulate the transfer of these changes from one generation to the next. By so defining the subject matter, I restricted the substance of the book to theories that began to emerge in the 1920's, after the birth of the science of genetics. I mention only briefly the older parts of evolutionary theory.

This book was originally intended as a reference text for students of life sciences. I attempted to write it in as simple a language as possible (without jeopardizing the scientific content) so that other potential readers with some elementary knowledge of biology and basic genetics – such as high-school biology teachers and pupils – should be able to understand it. On the other hand, I did not think it necessary to elaborate on topics of general zoology or genetics, which can be obtained from other sources. At the end of each chapter I have added a list of books and articles for supplementary reading.

In the mid-20th century, progress in mathematical population genetics and evolutionary biology ran fast ahead of the accumulation of data in support of the theory. My experience in teaching the subject convinced me that the ordinary biology student shrinks from mathematical formulations. But it is often easier to explain theoretical ideas using such formulations than to do so verbally. I have tried to find a balance between the theoretical, mathematical, and the biological presentation of evolutionary processes. I included some mathematical formulations – in basic algebraic form – and arranged them in "boxes", accompanied by numerical examples (from my own experiments, or taken from the literature). Readers who do not like, or do not need, these formulations may skip the boxes.

Many of the examples of natural selection which are included in this book, are based on observations of more than 50 years ago. It may seem old-fashioned and perhaps out-dated in this modern age of DNA, to use examples of morphologically-based studies. I can assure the reader that no better examples can be found in the molecular literature to illustrate these processes (I deal briefly with molecular evolution in chapter 20. An extensive discussion of this important subject can be found in Graur & Li, 2000). The student of evolution should constantly bear in mind that evolution does not occur within the nucleus or a cell. The fate of a mutation – whether it is fixed and survives or is lost – is decided in populations of organisms. The molecular genetic changes – mutations – within the cell can only affect evolution through the survival or mortality of the individuals in the cells of which they occurred. Random processes (genetic drift), which play an important role in evolution, matter only when a small number of organisms – perhaps a single one – are "drifted" somewhere. Natural selection is only able to detect the molecular changes if they are expressed in the phenotype of some individual organism. Thus, the morphological studies are as relevant today as they ever were.

I have arranged the material in the book in three parts. In Part 1, I present the basic quantitative theory of evolution, sprinkled with some examples, and with some of the algebraic formulations presented in "boxes". Part 2 opens with a historical review, and is dedicated to natural selection in natural populations ("micro-evolution"). Part 3 deals with macroevolution – speciation, extinction and phylogeny. The book ends with a discussion of human evolution and its impact on the biological world.

There is no doubt that the theory of evolution is the only scientific theory able to explain the known facts of biology. I have attempted not to smooth over difficulties of interpretation and scientific controversies, but to show that evolutionary theory is itself an evolving subject.

It is impossible to include in the book all the facts and theories which have been published on evolutionary processes. The selection of subjects to be included in the book, and the examples chosen to illustrate the processes, reflect my own personal preferences and not necessarily the most popular, or common, views of the subjects.

I hope that the book will contribute to the interest in, and knowledge of, evolutionary processes among biologists and the general public.

David Wool

PART I

Mainly Theory

CHAPTER 1

The Beginning

Is evolution a theory, a system, or a hypothesis? It is much more: it is a general postulate to which all theories, all hypotheses, all systems must satisfy in order to be thinkable and true. Evolution is a light which illuminates all facts, a trajectory which all lines of thought must follow. (Teilhard de Chardin, cited by Dobzhansky, 1961, p. 347)

Evolution, as a scientific discipline, comprises three main fields of research: 1) The origin of life, 2) Evolution as the description of the history of the biological world (the lines of descent connecting different groups of organisms), and 3) Evolution as an ongoing process. I shall deal with the first subject briefly in this chapter. Most of the material in the book deals with the third. The second subject (phylogeny) is deferred to the end of the book.

THE POINT OF DEPARTURE

The starting point of evolution as a science is the conviction that the biological world as we know it has evolved, gradually and slowly, from non-living matter.

There are those who, even today, do not accept this starting point, and believe that animals and plants were created by the Deity in six days, as written in the first chapter of Genesis. Such 'creationists' objected to the theory of evolution from the start, and continue to do so today. In some states of the USA, demands to remove evolution from the school curriculum, or alternatively, demanding 'equal time' for 'creation science' as an equivalent scientific theory to evolution, are brought up periodically. The dispute has recently (2004) entered the pages of the oldest, prestigious scientific journal, Nature. In an editorial (Nature 434: 1053) it was suggested that scientists should try to convince creationist readers that their belief can be reconciled with science, and that the journal should not reject creationist papers out of hand as unscientific (ironically, Nature was co-founded in 1869 by T.H. Huxley, a self-declared agnostic who fought fiercely against religious dogma!). The editorial evoked angry responses from many scientists.

Creationism is not a <u>scientific</u> equivalent to the theory of evolution. Creationists believe that the world was created and is controlled by the external power of the Deity, and, as such, the laws which regulate the world are beyond the grasp of the human mind. Scientists seek answers for questions about nature by studying nature itself, and derive explanations and suggest rules for its regulation from observations and experiments, not from external forces. The dispute between creationists and scientists is not about evolution: it is about the belief in God. There are no scientific tools to deal with matters of belief.

The 'proofs' and arguments brought forward by creationists to support their belief are not new. The same arguments and examples were published two hundred years ago by the British theologian, William Paley, in his book 'Natural Theology' (1802) (In fact, Paley seems to have had a much better knowledge of biology than some of the present-day creationists). The main argument is that complex structures like the human eye, so beautifully adjusted for its function as an organ of vision, could not have been formed by chance: it seems that the eye – and organs like it – were designed for the function they perform, and where there is a design – there must be a designer ('intelligent designer' (ID) is the most recent catch phrase for creationism). But presenting evolution as a matter of chance is wrong and misleading. <u>Evolutionists claim that the structures were formed not by chance, but rather by the gradual accumulation of improvements from simple structures</u>, which then survived because they had given some advantage to individuals over the carriers of simpler structures.

Among evolutionists, there is a diversity of opinions regarding some mechanisms of evolution (e.g., the importance of selection versus random processes, or whether evolution has been gradual or 'saltational'). The theory of evolution has evolved over the years, with changes in emphasis about such matters as the lines of descent among some groups of species (even the concept of a 'species' has been controversial). But all these approaches are disputes among scientific theories, starting from the same point of departure – and considering natural processes only. Creationism is not a scientific theory because it allows control by external forces, and it cannot be an alternative to the theory of evolution. As one of the greatest

British scientists, 'the father of Geology', Charles Lyell – wrote in the early 19th century:

> "For what we term "independent creation" or the direct intervention of the Supreme Cause, must simply be considered as an avowal that we deem the question to lie beyond the domain of science" (Lyell, 1873, "The Antiquity of Man. p. 468)

THE ORIGIN OF LIFE

Where? It is assumed almost universally among evolutionists that life began on planet Earth, at a remote time when conditions were suitable. Some, however, believe that life originated on some other star, and was transported to earth via meteorites. This alternative is not very useful for the understanding of the origin of life, since it only removes the problem to sites as yet unapproachable by contemporary science. Moreover, even if life had arrived from somewhere else, it must have been in a very primitive form, and we still need to have a theory to explain how the present diversity of living forms evolved. As T.H. Huxley wrote in 1870:

> "Whether they have originated in the globe itself, or whether they have been imported by, or in, meteorites from without, the problem of the origin of those successive faunas and floras of the earth… remains exactly where it was. For I think it will be admitted that the germs brought to us by meteorites, if any, were not ova of elephants, nor of crocodiles; not cocoa nuts nor acorns; not even shellfish and corals, but only those of the lowest animal and vegetable life. Therefore…the higher forms of animals and plants…either have been created, or they have arisen by evolution (T. H. Huxley, Biogenesis and Abiogenesis, 1870).

Spontaneous generation of life

In older times, people believed in the formation of living organisms from non-living matter: fish and frogs were born from the mud in the bottom of streams and ponds, etc. Van Helmont, as late as the 17th century, offered the following recipe for the formation of 'artificial' mice:

> Close the mouth of a jar containing wheat grains with a dirty shirt. After about 20 days, a ferment from the shirt combines with materials from the wheat, and the grains become mice of both sexes, which can mate among themselves or with other mice. These mice…are perfect and do not have to be suckled by their mother (cited in Glass et al., 'Forerunners of Darwin', p. 40).

The belief in spontaneous generation of organisms from non-living matter was proved wrong as early as 1668. The Italian Francesco Redi put carcasses of a snake, fish, eel and a piece of meat in four jars, and sealed the jars with paper. Four other jars with similar contents were left unsealed.

Maggots appeared shortly in the open jars, but not in the sealed ones. This proved that maggots were not formed spontaneously from the rotting meat. Redi then repeated the experiment, sealing the jars with a thin net instead of paper. The results remained unchanged, but he noticed where the maggots came from: flies hovered around the jars, and maggots hatched from eggs the flies laid. Unable to penetrate the jars, the maggots died on the net.

With the improvement of the microscope, the problem of spontaneous generation emerged again: minute creatures (animalcules) were seen swarming in every drop of water. Surely these were formed spontaneously?

Jean Baptiste Lamarck, at the Museum of Natural History in Paris, strongly argued the case. He suggested that these animalcules were the only forms of life which Nature could create spontaneously from non-living material. All other forms were descended from this source:

> Nature, by means of heat, light, electricity and moisture, forms direct or spontaneous generations at that extremity of each kingdom of living bodies, where the simplest of these bodies are found (Lamarck, Philosophical Zoology 1809 (1917), p. 244).

This possibility was also proved wrong. Another Italian, Spallanzani, about 50 years before Lamarck, boiled a broth and then sealed the necks of the flasks: no animalcules – known today as bacteria and yeast – appeared in the sealed flasks. This result was confirmed in the mid-19th century by Louis Pasteur. The method of sterilizing foodstuffs ('pasteurization') has been in use since then for protection and preservation of commodities. Sterilization procedures were introduced by Lister to contain bacterial infections in British medical institutions, and saved countless lives.

Even these simple germs ('monads', as Lamarck called them), thus, do not form spontaneously: our predecessors only assumed they did because it had not yet been discovered how these organisms reproduced –

> "We must be on our guard not to tread in the footsteps of the naturalists of the Middle Ages, who believed in spontaneous generation to be applicable to all those parts of the animal and vegetable kingdoms which they least understood…and when at length they found that insects and cryptogamous plants were also propagated from eggs or seed, still persisted in retaining their prejudices respecting infusory animals, the generation of which had not been demonstrated by the microscope to be governed by the same laws" (Lyell, Principles of Geology, 1853, p. 580).

But if so, how did life begin?

> "I admit that spontaneous generation, in spite of all the vain attempts to demonstrate it, remains for me a logical necessity" (A. Weismann, 1881, p. 34-35).

Modern approaches

In the 1940s, a few laboratory experiments opened the way for the understanding of the origin of life. The experiments showed that organic molecules – among them amino acids and even short peptides – can be formed from inorganic materials like methane (CH_4), ammonia (NH_3) and water vapor, when an external source of energy is applied. Ultraviolet light, electric sparks and other forms of energy have been successfully used (Miller,1953). There is evidence that the atmosphere of the primitive earth may have contained these gases (Miller, 1974).

The hypothesis is that life began in a mixture of organic molecules, a **'primeval soup'**. Interestingly, Darwin had the same idea when he mused on the question why we are not likely to see around us newly-formed living material:

> "If we could conceive, in some warm little pond, with all sorts of ammonia and phosphoric salts, light, electricity (&c) present, that a proteine [sic] compound was chemically formed ready to undergo still more complex changes, at the present day such matter would be instantly devoured or absorbed, which would not have been the case before living organisms were formed" (Darwin, Letters III: p.18).

But the 'primeval soup' was still far from containing even the most primitive living organism, such as a bacterial cell. Several hypotheses were offered to account for the next stage – the agglomeration of different molecules to form a structure. A critical point for these hypotheses is that some self-replicating molecules were formed. As is known today, self-replicating molecules are DNA and RNA, which carry the genetic information. However, DNA molecules do not form spontaneously: for their formation, the activity of enzymes is required – and enzymes are proteins, which need DNA as a template for their formation. So we have a 'chicken-and-egg' problem: Which came first and how?

Scientists interested in the origin of life have been trying to suggest theoretical scenarios which overcome this difficulty (e.g., Lahav, 1993). All these models assume some selective advantage to particular molecules or combination of molecules in the primeval soup. So far there is little evidence in support of these theories. The origin of life thus remains obscure.

The customary model is that there was a lapse of $3.5*10^9$ years from the origin of the earth to the appearance of the first cells, followed by a gradual evolution of the early multicellular organisms which was completed in a short span of $5*10^6$ years. It should be mentioned that the scenario outlined above, of gradual development in the primeval soup, is not the only possible scenario for the early evolution of organisms. Schwabe (2002) suggested that the genomes of all extant organisms – from yeast to man – evolved simultaneously, so that proto-cells with genetic potentialities ready to develop into all biological classes were available 3.5 billion years ago (the 'genomic potential' hypothesis). The 'genomic potential' is considered a function of the chemical affinities of the molecules, not a result of competition and adaptation (Schwabe, 2002). This hypothesis explains why representatives of each class of organisms – when they first appear as fossils in the palaeontological record – are complete with all their appendages and (presumably) senses, and why sequence differences of cytochromes and other proteins in organisms as different as yeast and man, do not show a linear order of divergence. However, this hypothesis does not seem to have many supporters.

Literature Cited

Dobzhansky, T. 1961. Mankind Evolving. Yale University Press, New Haven, USA.

Glass, B., O. Temkin and W.L. Strauss. 1968. Forerunners of Darwin. Johns Hopkins Press, Baltimore, USA.

Huxley, T.H. 1881 (1890). Biogenesis and abiogenesis. Critiques and Addresses, MacMillan & Co., London, UK. Pp. 218-250.

Lahav, N. 1993. The RNA world and co-evolution hypothesis and the origin of life: implications, research strategies and perspectives. Origins of Life and Evolution of the Biosphere 23: 329-344.

Lamarck, J.B. 1917 (1809). Zoological Philosophy. MacMillan & Co., London, UK.

Lyell, C. 1853 (1830-1832). Principles of Geology, 9th edition. Murray, London, UK.

Lyell, C. 1873. The Antiquity of Man. Murray, London, UK.

Miller, S. 1953. A production of amino acids under possible primitive earth conditions. Science 117: 528-529.

Miller, S.L. 1974. The atmosphere of primitive earth and the prebiotic synthesis of amino acids. Origins Life 5: 139-151.

Paley, W. 1802 (1844). Natural theology, or evidences of the existence and attributes of the Deity, collected from the appearances of nature. Paley's Works, H.G. Bohm, London, UK.

Schwabe, S. 2002. Genomic potential hypothesis of evolution: a concept of biogenesis in habitable spaces of the universe. The Anatomical Record 268: 171-179.

CHAPTER 2

Evolution as an On-going Process

It is generally accepted that evolutionary changes are slow. This concept was crucial for Darwin's ideas about natural selection (the idea that vast amounts of time were required was advocated by Charles Lyell in his 'Principles of Geology', which Darwin read during his voyage on the Beagle). If our conception of evolution is correct, the same kinds of processes that drove evolution in the past should be going on in our own time, just as the geological forces of erosion, volcanic eruptions, and tectonic changes continue. The rates of these processes may vary from one organism to another and in different locations. We are usually unable to detect these changes, either for lack of suitable markers or because they are too slow relative to our lifetime. But some such changes are fast enough to be detectable, and may help us to understand those that are not.

We define an evolutionary process as any process that causes a permanent, heritable (genetic) change in composition of a population. This change may be the result of natural or artificial selection, like the widespread phenomenon of insecticide resistance in insects; but it could be brought about by other means as well. For example, when a population is founded by a single seed or a single fertilized female landing at a new site, its future genetic composition may be strongly affected by chance.

The final result of an evolutionary process depends on the direction and intensity of the factors that bring it about. A short and ephemeral environmental alteration may not have an effect, but a long-lasting environmental change may cause a permanent loss of some of the gene pool of the population. Immigration from or hybridization with a neighboring population may add new genes to it. These changes may or may not be expressed in the morphological appearance of the organisms, but with time they may accumulate and lead to a reproductive barrier, the first step in the formation of a new species.

BASIC CONCEPTS AND DEFINITIONS

Population

A population is an assemblage of individuals. The term population is given different meanings in everyday use, not always in agreement with its biological meaning. For example, the 'population' occupying a coral reef includes very different organisms – fishes, corals, sea urchins, worms and algae – belonging to different biological classes. Changes in the genetic composition of one of these groups may – but not necessarily will – affect other groups. Biologically, these are different populations which occupy the same site.

The definition of a biological population in the context of population genetics and evolution requires that <u>all individuals in it belong to one and the same biological species, and live at a defined space (site) and time.</u> Genetically, a population consists of (potentially) interbreeding individuals: genes can be freely interchanged among them. Populations vary with time and from one site to another. Research may focus on variation in space, or follow a single population as it changes with time.

There is a certain resemblance between a population and an individual organism. An organism is born, grows, matures, ages and dies. A population is 'born' when founded, often by a small number of individuals, and grows by their reproductive efforts. It ages as the proportion of individuals past reproductive age increases and the proportion of newborn in a unit of time decreases. The population dies when births stop or are insufficient to balance the deaths.

But this comparison fails to expose some important properties of populations, which are not present in individuals. The genetic potential of an individual is determined at the time of fertilization of the egg cell by a sperm (in sexually reproducing organisms) and does not change throughout life. The genetic composition of a population begins with a pool of genetic material carried by the founders, and keeps changing as new individuals are born, immigrate into it, emigrate from it, or die (a local population may become extinct (die) even when all the individuals composing it remain alive – as when they move to another site). The gene pool may change when external forces operate. These are the **evolutionary changes**, which are the focus of this book.

GENETIC VARIATION IN POPULATIONS

Natural populations are rarely homogeneous (asexually-reproducing organisms are an exception). The number of genotypes in a population of sexually-reproducing organisms may be equal to the number of individuals in that population: each individual may be genetically unique.

To simplify the discussion, let us suppose that we are dealing with a diploid organism which has only a single gene, with two alternative alleles, A, a. [I assume that the reader is familiar with the basic genetic terminology.] There are three possible diploid genotypes: AA, Aa, and aa.

A more complex organism with 2 genes, each with 2 alleles, can produce 9 genotypes ($= 3^2$): aabb, aaBb, aaBB, Aabb, AaBb, AaBB, AAbb, AABb, AABB. If the number of alleles per gene is **k**, the number of possible combinations is $k*(k + 1)/2$, and for **n** genes the number is $(k*(k + 1)/2)^n$. Even in a simple organism like the fruit fly *Drosophila melanogaster,* which has only 4 pairs of chromosomes (humans have 23), **n** is estimated as several thousand genes and the number of possible combinations becomes astronomical. The probability that two organisms in a population are genetically identical is very small, except in the case of twins, or when the population is inbred or reproduces asexually. These exceptions will be dealt with separately.

GENETIC AND PHENOTYPIC VARIATION

The enormous amount of genetic variation in populations does not seem to be reflected in what we observe around us, in any population of any organism. All the sparrows around our dwellings look similar to us. We certainly cannot tell one housefly apart from another. But there is no doubt that other organisms, e.g., sparrows, recognize other individuals of their species (they may find it difficult to tell apart two humans!).

A basic reason for our inability to tell individuals of other species apart – while we have no problem in identifying other humans individually – is genetic. The **genotype** of any individual is only partly expressed in the **phenotype** (external appearance). Many genes that are carried by an organism may be expressed in characters that our eyes, or other senses, cannot detect – or may not be expressed at all (a large part of the DNA in the nucleus appears not to be expressed – 'junk DNA', see Graur and Li, 2000). These genes can be identified using sophisticated molecular techniques, which are becoming more and more common but are still not easy to use. The older techniques of serology and protein electrophoresis are more straightforward and easy to understand and are still being used. All these methods support the conclusion that individuals are in fact different from each other genetically (DNA 'fingerprinting' is used to identify individuals in forensic and criminal investigations and in paternity disputes in humans and in birds).

It should be remembered that the phenotype – but not the genotype! – of individuals is affected directly by the environment. Body size, for example, is strongly affected by food supply, in particular in the larval stages of insects. The body color of a chameleon changes in response to the physical characteristics of the surface on which it finds itself – and the animal's 'mood'. Genetic phenomena like dominance or incomplete penetrance (e.g., Wool and Mendlinger, 1973) reduce the observed phenotypic variation because not all individuals that carry a gene also express it in the phenotype. These phenomena complicate the quantitative estimation of genetic variation in a population.

HOW TO DETECT EVOLUTIONARY CHANGE

How can we know that an evolutionary (genetic) change has occurred in a population? How can the current information on the genetic composition of the population, and on the forces that may affect it, be used to predict the direction and the magnitude of changes that may occur in the future?

This is a complex question, and the answer is not simple. In controlled mating systems, the proportions of genotypes in the progeny can be predicted from knowledge of the genotypes of the parents (provided that we are dealing with a single or only a few heritable characters). The Czech monk Gregor Mendel showed, as early as 1865, that if you cross a red-flowered pea variety (RR) with a white-flowered one (rr), you get in the first generation all-red flowered seedlings (Red is dominant to white. All offspring are in fact hybrids, Rr) and in the second generation you get red- and white-flowered seedlings in the ratio 3:1 (the genotypes AA, Aa and aa are expected to be in the ratios 1:2:1). Mendel explained his results by two laws - the law of segregation and the law of independent assortment - which became known to the world 35 years later when his paper was rediscovered in 1900. These laws are the foundation of the science of genetics and are described in many elementary textbooks. It should perhaps be emphasized that these predicted ratios do not mean that every seed pod, or the yield from any individual plant, will contain exactly these proportions of genotypes: these are probabilities. The actual numbers produced will depend, among other things, on sampling variation and the number of offspring produced per pair.

When dealing with a population, outside the limited cases of controlled mating, prediction becomes much more complex. Even if we do know the genotypic proportions in the parents, we will not be able to predict their proportions in the offspring population unless we also know the <u>frequency of matings</u> between each pair of genotypes, and the <u>reproductive output</u> of each genotype combination.

Even in the simple case of one gene and 2 alleles, in a freely mating population, there are 9 possible combinations of genotypes (AA × AA, AA × Aa, … aa × aa). Each combination yields offspring in different proportions (expected from Mendel's laws), and different crosses most probably differ in the number of progeny. We have little data on genotype frequencies and mating preferences in any natural population. It is unlikely that we shall ever acquire all the information necessary in order to predict the genotype frequencies in any population. The best we can do is to estimate the proportions of progeny in terms of probabilities based on theoretical considerations of an <u>ideal</u> situation – in other words, to use a **model**.

The model will necessarily be constructed for the simplest cases. We will need a set of easily detectable **genetic markers** in each population – markers that can be conveniently followed and yet be representative of the genome as a whole. (This last is a difficult point, because we rarely know how closely the marker responds to the processes that affect other genes.)

MODELS IN POPULATION GENETICS AND EVOLUTION

In order to understand the complex real world, we must simplify the phenomena and try to explain them using the smallest possible number of factors. In doing so, we necessarily ignore other factors, which may have important effects.

Almost all the theoretical studies deal with **models** of simple systems of only a few genes and alleles. The models are further simplified by assuming that the genotype is identical with the phenotype, i.e. that phenotypic changes reflect accurately the genetic changes in the population.

It is not difficult to realize that <u>processes in nature do not work on isolated genes but on entire organisms,</u> which carry thousands of genes. When we cross a red-eyed *Drosophila* male with a white-eyed female we are mixing two genomes. The eye color is only a genetic marker, enabling the researcher to follow the results of the cross – but the mixing may have other, undetected effects (Franklin and Lewontin, 1970).

A model is good if it is useful – allowing the researcher to predict an outcome. The prediction can be tested and verified – or rejected – by observation and/or experiment. To make a model more 'realistic', more and more factors may have to be incorporated in it, but the model may then become too complex to be useful. There is no use for a model which is as complex as nature itself.

The simple models in population genetics and evolution enable the formulation of 'working hypotheses', which can be experimentally verified. The most important model in population genetics theory is the **Hardy-Weinberg equilibrium law**.

Data on the composition of the population at a given point in time (as estimated by the frequencies of genotypes at a marker locus), are used to predict what would be its composition at the next point (generation) in time, provided no external force is affecting it. The population is then at a **genetic equilibrium**. The predicted frequencies are compared with actual, observed frequencies. Discrepancies between prediction from the 'ideal' model and reality may indicate that an evolutionary change has taken place.

Literature Cited

Franklin, I. and R.C. Lewontin. 1970. Is the gene the unit of selection? Genetics 65: 707-734.

Graur, D. and W-H. Li. 2000. Fundamentals of Molecular Evolution. Sinauer, Sunderland, USA.

Wool, D. and S. Mendlinger. 1973. The eu mutant of the flour beetle, *Tribolium castaneum* Herbst: environmental and genetic effects on penetrance. Genetica 44: 496-504.

CHAPTER 3

Populations at Equilibrium:
The Hardy-Weinberg Law

The Hardy-Weinberg Law is a model which predicts the genotype frequencies at a marker locus in generation **n + 1**, from their frequencies at generation **n**, provided that no external force is operating that may change the allele frequencies.

The model describes a single, independent marker locus with two alternative alleles, **A, a.** The genotypes at the locus are expressed completely in the phenotypes of the organisms, which can therefore be recognized and counted.

Definitions.   In a population with N diploid individuals, the total number of alleles (**A** and **a**) is 2N. This total is referred to as the **gene pool** of the population. Each **AA** individual contributes 2 alleles **A** to the pool; each **aa** individual, 2 alleles **a**, and an **Aa** individual one of each. If a sample of N contains D* individuals of **AA**, H* individuals of **Aa**, and R* individuals of **aa** (**D* + H* + R* = N**), it is customary to define the **genotype frequencies** as D = D*/N, H = H*/N, and R = R*/N, respectively, so that D + H + R = 1.0.

From these frequencies we calculate **allele frequencies**:

$$P_{(A)} = D + \tfrac{1}{2} H$$

$$q_{(a)} = R + \tfrac{1}{2} H$$

then p + q = 1.0

GENETIC EQUILIBRIUM

A population is at equilibrium if its allele frequencies p, q remain constant and do not change with time (mathematically, $p_{n+1} = p_n$ and $q_{n+1} = q_n$).

In 1908, G.H. Hardy in England and W. Weinberg in Germany independently derived the conditions required for a genetic equilibrium. Beginning from any combination of genotypes that sum to 1 (D + H + R = 1.0), the population will, in one generation, reach the following equilibrium frequencies

$$\text{Frequency of AA} = p^2$$

$$\text{Frequency of Aa} = 2pq$$

$$\text{Frequency of aa} = q^2$$

The equilibrium frequencies are derived from the expansion of the binomial,

$$(p + q)^2$$

These frequencies will remain stable and not change as long as the following **requirements** are met:

(1) The population is infinitely large.

(2) Mating in the population is random

(3) There are no mutations (at the observed locus)

(4) There is no migration in or out of the population

(5) There is no selection.

The calculations of equilibrium frequencies are illustrated in Box 3.1. This is done in two steps. We start with any genotype frequencies D, H, R.

First we calculate the frequencies of mating between each pair of genotypes (Box 3.1, A). In a random mating population (requirement 2), the probability of mating between two genotypes is the product of their frequencies in the population. We then proceed to list the offspring of each type of mating from Mendel's laws: e.g., the offspring of mating **Aa** × **AA** will be 50% **AA** and 50% **Aa** (Box 3.1 B). The offspring are then summed by genotype, which yields the expected equilibrium frequencies (bottom of the box).

If we start with a population at equilibrium, the final sums will remain the same (the reader may replace D, H, R in Box 3.1 by p^2, $2pq$, q^2 respectively, and repeat the calculations).

Box 3.1	Calculation of expected equilibrium frequencies

A. Mating frequencies

		Male genotype		
Female genotypes		**AA**	**Aa**	**aa**
	(frequencies)	(D)	(H)	(R)
AA	(D)	D^2	DH	DR
Aa	(H)	DH	H^2	HR
aa	(R)	DR	HR	R^2

B. Offspring frequencies

Parent genotypes	mating frequencies	offspring		
		AA	Aa	aa
AA × **AA**	D^2	D^2		
AA × **Aa**	2DH	DH	DH	
Aa × **Aa**	H^2	$\frac{1}{4} H^2$	$\frac{1}{2} H^2$	$\frac{1}{4} H^2$
AA × **aa**	2DR		2DR	
Aa × **aa**	2HR		HR	HR
aa × **aa**	R^2			R^2

Offspring totals: **AA:** $D^2 + DH + \frac{1}{4} H^2 = (D + \frac{1}{2} H)^2 = p^2$

Aa: $2 (D + \frac{1}{2} H) (R + \frac{1}{2} H) = 2pq$

aa: $R^2 + RH + \frac{1}{4} H^2 = (R + \frac{1}{2} H)^2 = q^2$

When the numbers of alleles at a locus exceeds 2, expected genotype frequencies can be calculated in a similar manner. For example, for three alleles at frequencies p, q, r, the frequencies of the six offspring genotypes can be calculated as $(p + q + r)^2 = p^2 + q^2 + r^2 + 2pq + 2pr + 2qr$.

The Hardy-Weinberg law describes an <u>unstable equilibrium</u>. Imagine a ball placed on a horizontal flat surface, like a pool (billiard) table (Fig. 3.1 B) At every point on the surface the forces of gravity and friction are balanced and the ball will remain where it was placed: it can be said to be in equilibrium. If a force is applied, it will move the ball until it settles at another equilibrium point: it will not return to the point of origin. In an analogous manner, if the genetic equilibrium of a population is disturbed (see in the following chapters), it will reach Hardy-Weinberg equilibrium at some other allele frequencies, and will not return to the original ones. <u>Genetic changes are not reversible</u>. This is the essence of the evolutionary process.

GRAPHIC ILLUSTRATIONS: DE-FINETTI DIAGRAMS

The composition of a population with 3 genotypes at given frequencies can be described as a point in triangular space. This method, suggested in

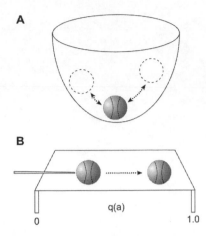

Fig. 3.1 Stable (A) and unstable (B) equilibria. A. If disturbed, the ball will come back to rest at the starting point. B. If disturbed, the ball will come to rest at a different equilibrium point.

1927 by De-Finetti (Spiess, 1977), is based on the geometric rule that <u>the sum of the lengths of perpendiculars from a point inside an equilateral triangle to the three sides of the triangle equals the height of the triangle</u> (see proof in Spiess, 1977).

The population composition is represented by the crossing point of the three perpendiculars, labeled D, H and R, in an equilateral triangle of unit length (Fig. 3.2. A point D% from the side labeled AA will lie on a line parallel to that side, D mm from it). As an example, the reader may position a population composed of 3 genotypes at frequencies 0.25, 0.5, 0.25 as the crossing point of the two arrows in the figure, using the grades on the opposite sides to position the arrows. Only two frequencies are needed to position a point since $D + H + R = 1$.

Properties of the De-Finetti Diagram

(A) The perpendicular from a point in the triangle to the base (H in Fig. 3.2), divides the base BC in two segments in the ratio q/p, as shown in the figure. Allele frequencies at the point can thus be read directly from the diagram.

(B) All equilibrium genotype frequencies for the entire range of q, from zero to one, form a parabola which spans the base of the triangle, with a maximum at $p = q = 0.5$ (this is the crossing point of the two arrows in Fig. 3.2).

This parabola is derived from the quadratic expression at equilibrium,

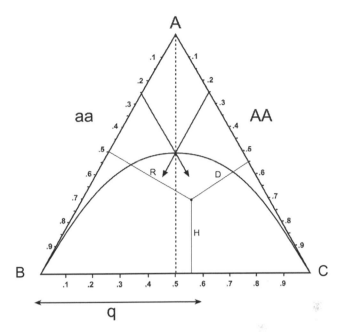

Fig. 3.2 A De-Finetti diagram. A population is plotted as a point in the triangle. The perpendicular distances from the point to the three sides of the equilateral triangle – D, H and R – represent genotype frequencies of **AA**, **Aa** and **aa** in the population. The point lies D% from side AC and R% from side AB (only 2 of the 3 frequencies are needed to plot a point in the triangle). All points D% from side AC (representing **AA**) – lie on a line parallel to that side, and similarly for R% (representing **aa**) – as illustrated by the two arrows in the figure. The parabola is a plot of all equilibrium frequencies. Note that the cross point of the arrows is 0.25, 0.5, 0.25 – an equilibrium point for **p** = **q** = 0.5, which is at the maximum height of the parabola.

$$H = 2pq = 2p\,(1-p) = 2p - 2p^2$$

which differentiates into dH/dp = 2 – 4p. At the maximum, dH/dp = 0 and therefore

$$4p = 2 \text{ or } p = 0.5$$

(C) De-Finetti diagrams may be used to illustrate population genetic trends over time by plotting genotype frequencies of a single population at sequential generations, as in Fig. 3.3, or to illustrate spatial variation among populations, by plotting several populations in the same triangle.

USES OF THE HARDY-WEINBERG LAW

The Hardy-Weinberg model enables the detection of an evolutionary process when the observed genotype frequencies in a population do not

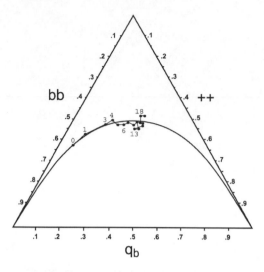

Fig. 3.3 Changes in the frequency of the **b** allele in a population of flour beetles initially founded at q_b = 0.25 and maintained for 20 generations. The allele frequency increased gradually with time, although at each generation the adult genotype frequencies did not depart significantly from equilibrium (after Sokal and Sonleitner, 1968).

conform to equilibrium expectation. Deviations from expectation may arise when any one of the five requirements of the model is not met. <u>The deviation from expectation in itself does not indicate which requirement is violated.</u>

Sampling error

Deviations from expectation can result from sampling error, even when all requirements of the model are met. Statistical tests are used to confirm that the deviation is unlikely to be due to chance. The most common are chi-square tests of goodness of fit (χ^2 tests. Consult any book on statistics, e.g. Sokal and Rohlf, 1995). If the observed frequency of a genotype is denoted O and the expected frequency E (numbers, not proportions) then the chi-square statistic is computed as

$$x^2 = \Sigma((O - E)^2/E)$$

The summation is over all genotypes (3 in the simple model). If the calculated value of the statistic exceeds that of the expected values (which can be obtained from an appropriate chi-square table), the assumption that the population is at equilibrium can be safely rejected (with the probability of making a wrong decision being less than 5%). No conclusions can be seriously derived from a comparison of observed and expected frequencies unless tested in this way.

The following numerical examples were taken from my own class experiments with flour beetles (*Tribolium castaneum*, Tenebrionidae) (Box 3.2). In part A the population seemed to be in equilibrium, not so in part B. The marker locus in both cases was the black-body mutation, **b**.

Box 3.2

A. Parental population: 9 pairs of **bb**, one pair of wild type (++) beetles (the frequency of **b** = 0.9)

Offspring genotypes:	**bb**	+b	++	Total
Frequencies observed:	295	80	6	381
Expected frequencies:	$p^2 = 0.81$	$2pq = 0.18$	$q^2 = 0.01$	
Numbers	308.61	68.58	3.81	381

$X^2 = 3.761$, p > 0.05. The null hypothesis of equilibrium cannot be rejected.

B. Parental population: One pair **bb**, 9 pairs wild type (++) beetles (frequency of **b** = 0.1)

Offspring genotypes:	**bb**	+b	++	Total
Frequencies observed:	2	260	372	634
Expected frequencies:	$p^2 = 0.01$	$2pq = 0.18$	$q^2 = 0.81$	
Numbers:	6.34	114.12	513.54	634

$X^2 = 228.64$, p << 0.001. Offspring frequencies deviate strongly from expected ones. The population is not in equilibrium.

Estimating the frequency of recessive alleles

When an allele is recessive, genotype counts do not reveal how frequent it is in the population: many of the carriers will be heterozygotes, in which the character is not expressed in the phenotype. Many genetic diseases in humans are due to recessive alleles. The carriers are healthy and do not suffer from the symptoms of the disease. The disadvantage is that they may pass on the defect to their children. The probability that some of the children will have the disease increases if they marry other heterozygotes, who are similarly unaware that they carry the disease. It is very important, e.g., for genetic counseling, to be able to estimate the allele frequency of such diseases in order to assess the risk.

In the absence of data to the contrary, scientists may consider that the population is at equilibrium. In this case, the observed frequency of affected individuals (preferably in a large sample) will be $R = q^2$, and the frequency of the recessive allele is $q = \sqrt{R}$. This is of course an estimate, which may be way off if the population is in fact not in equilibrium – but in many cases it is the only possible way to estimate the frequency of carriers ($H = 2pq$, when $p = 1 - q$).

For example, the frequency of the recesive allele *eu* in experimental populations of the flour beetle *Tribolium castaneum* was estimated in order to monitor the efficacy of natural selection against it. When homozygous, this allele causes the appearance of extra appendages in the larva and pupa (Wool and Mendlinger, 1973). The initial population was a cross **eu/eu** x +/+, yielding all heterozygous progeny ($q_0 = 0.5$) (Table 3.1).

Table 3.1 Changes in the frequency of the eu allele

Generation	Total counted	# mutants	% mutants (R)	$q = \sqrt{R}$
1	1327	0		0.5
2	1726	147	0.085	0.291
3	1101	63	0.057	0.239
4	1465	90	0.061	0.247
5	1590	69	0.043	0.207
6	1557	67	0.043	0.207
7	1614	62	0.038	0.195

DEVIATIONS FROM EQUILIBRIUM: GRAPHIC METHODS

Absolute Limits

This method is suitable for experimental insect populations. When the initial number of eggs in the experimental replicate is known but the numbers emerging as classifiable adults are smaller, as is often the case, or when not all the adults can be classified due to heavy work load, the following method may help decide whether the population is at equilibrium or not. Calculate the observed and expected offspring genotype frequencies from the available sample and plot them in a De-Finetti diagram. Then add the numbers of underlined individuals in turn to the observed AA, Aa and aa numbers and recalculate the genotype frequencies. Plot the new points in the figure around the expected frequencies (as in Fig. 3.4). The triangle thus formed contains all the possible genotype combinations (had classification been complete) and constitutes the absolute limits around the observed point. If the expected point remains outside these limits, accidental mortality or sampling cannot be the reason for the deviation from expectation.

The Hexagon Method

This method is useful if several samples are taken from the same population, and deviations from expectation in some of the samples are

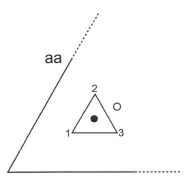

Fig. 3.4 Absolute limits for accidental deviation from expectation. The observed frequencies are represented by the filled circle. The manner of calculating the limits is explained in the text. The expected point (white circle) is outside the absolute limits, therefore the population is not in equilibrium.

detectable by chi-square tests. One may wish to know if different samples show the same pattern of deviations (see below). If they do, the reason for the departure from equilibrium may be easier to work out or to ascertain. The hexagon method was suggested by Jolicoer and Brunel in 1966, and was successfully adopted by Sokal and Sonleitner (1968) and Wool (1970) (Fig. 3.5).

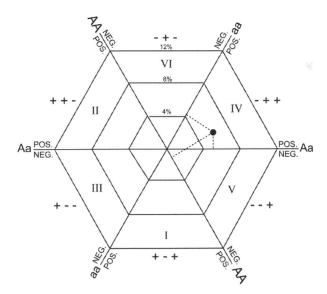

Fig. 3.5 The hexagon method for illustrating patterns of deviation from equilibrium in samples from the same population. Each component triangle represents one of the six possible patterns of deviation, as marked in the figure. For details see text.

The observed frequencies of each genotype may be higher (+) or be lower (−) than expected. This results in six possible patterns of departure from expectation, represented by the six equilateral triangles composing the hexagon and labeled − + − ; + + − ; + − −; − + + ; − − + ; + − + (the order of the symbols in each triplet is for genotypes **AA, Aa, aa** respectively).

Concentric hexagons are plotted on the same triangular graph paper used for De-Finetti diagrams (Fig. 3.5). The center of the hexagon − where the three diagonals, labeled **AA, Aa, aa** cross − represents the equilibrium proportions for that marker (zero deviations). <u>The data plotted in the hexagon are the differences (in percent, with the sign of the difference preserved) of the observed from expected frequencies of the three genotypes</u>: $(AA) = D − p^2$; $(Aa) = H − 2pq$; $(aa) = R − q^2$. Deviations may be positive (POS) or negative (NEG) and plotted on the appropriate sides of the diagonals. A sample with a deficiency of **AA** and an excess of **Aa** and **aa** is located in section IV in the hexagon (Fig. 3.5). An accumulation of sample points in any one section of the hexagon indicates a common pattern of deviations from equilibrium.

Cases where the Null Hypothesis of Equilibrium cannot be Rejected

A population may seem to be in genetic equilibrium − the observed genotype frequencies may not differ from expectation − and yet be affected by very strong selection processes.

The phenomenon described by Sokal and Sonleitner (1968) is particularly instructive. In their experiment with flour beetles, <u>genotype</u> frequencies of the adults − estimated every generation − rarely deviated from expectation, as illustrated by their proximity to the parabola in Fig. 3.3. But surprisingly, in lines started with low marker <u>allele</u> frequency, $q_b = 0.25$ (less so in the $q_b = 0.5$ lines), the frequency of **b** increased dramatically with time. The movement to the right of sequential generation points in Fig. 3.3 indicates that some force was acting to move the frequency of **b** away from the previous equilibrium point!

This apparent contradiction was explained by strong selection in the larval and pupal stages, which cancelled each other by the time the immatures became adults.

MORE COMPLEX MODELS

When we deal with a single marker locus, a population will reach Hardy-Weinberg equilibrium genotype proportions in one generation, provided that the five requirements are met. However, genes are not independent - for one thing, they are arranged in linkage groups on chromosomes. How quickly equilibrium may be reached when more than one locus is observed?

Two independent loci

We examine the case of two marker loci, **A**, **B**, each with 2 alleles. As before, we assume that the population is infinitely large, mating is random, and there are no mutations, migration or selection affecting these loci. Instead of three, we now have nine genotypes to monitor (Box 3.3). The computations become more tedious when more genotypes are involved (if we replace the genotype codes by their frequencies in the population, the three row sums will be the frequencies of the genotypes at locus **A**, and the three column sums – those of locus **B**.)

Box 3.3	Calculations of equilibrium frequencies at two independent loci

Genotypes:

	AABB	AABb	Aabb
	AaBB	AaBb	Aabb
	aaBB	aaBb	aabb

Gametic combinations:

	p(A)	q(a)
u(B)	pu	qu
v(b)	pv	qv

Genotype proportions at equilibrium:

p^2u^2	$2p^2uv$	p^2v^2
$2pqu^2$	$2pquv$	$2pqv^2$
q^2u^2	$2q^2uv$	q^2v^2

The nine genotypes produce four types of gametes – **AB, Ab, aB, ab** – (an alternative, equivalent expression of the fact is that the nine genotypes are produced by combinations of the four types of gametes). If the genes are independent, then each of the two genes will have equilibrium frequencies (sum the rows and columns to get p^2, $2pq$, q^2 for gene **A**, and u^2, $2uv$ and v^2 for gene **B**).

Continuing the calculations for the next generation involves 81 mating combinations. Unlike the single locus system, if we start with genotype proportions which are not in equilibrium, equilibrium is not reached in one generation even when all Hardy-Weinberg requirements are met! The approach to equilibrium may be slow, and will depend on the frequencies of **AB** and **ab** versus **Ab** and **aB** gametes. If we define

$$d = (pu * qv) - (pv * qu)$$

then at equilibrium **d** must be zero, or **pu*qv = pv*qu**. The approach to this stage may be slow, and **d** will be halved each generation:

$$d_n = (½)^n d_0$$

when d_0 is the initial value of **d**.

When the genes are linked, the approach to equilibrium will be even slower. If r is the frequency of recombination between genes **A** and **B**, and **c** is the proportion of non-recombinants ($c = 1 - r$) then

$$d_n = c * d_{n-1}$$

(in the special case of independent genes (no linkage), $c = \frac{1}{2}$.

Sex-linked Genes

Another case of complexity in calculating equilibrium frequencies concerns sex-linked genes. In many diploid organisms, males and females differ in one pair of chromosomes, referred to as the **x-y** system. The homogametic sex – in most cases, the female – has two **x**- chromosomes, and the heterogametic sex – commonly the male – carries one of each. Females therefore produce only **x**-carrying gametes, while males produce **x**- and **y**-carrying gametes in equal proportions.

It was shown by T.H. Morgan early in the 20th century that in sex-linked genes like the white-eye (w) gene of *Drosophila melanogaster,* reciprocal crosses do not give identical results. Recessive sex-linked characters will always be expressed in the males, but in females they will only be expressed when homozygous. Since the male receives his only **x** chromosome from his mother (he gets **y** from his father), a sex-linked trait is always transmitted from mother to son. Daughters are carriers of the trait and may not show it phenotypically, but transmit it to their own sons (Fig. 3.6).

Well-known sex-linked genes in man are those for color-blindness and haemophilia. These genetic defects are detected much more frequently in men than in women, and are sometimes thought to be specific to the male sex, but homozygous women will express them as well.

Equilibrium Frequencies in Sex-linked Genes

In a population, the frequencies of sex-linked alleles phenotypically expressed in males should be higher than in females: if the frequency of a sex-linked allele is q, then its frequency among male phenotypes will also be q, while in females it will be q^2. Since q is smaller than 1, $q^2 \ll q$ (e.g., if $q = 0.01$, then $q^2 = 0.0001$). The genotypic frequencies of a sex- linked allele in a sample of males should be a good estimate of its frequency in the population as a whole.

The ratio of the phenotypic frequencies in males should be higher than in females by approximately the factor $q/q^2 = 1/q$.

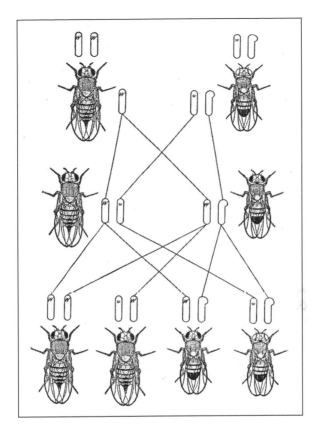

Fig. 3.6 Sex-linked inheritance in *Drosophila*: the cross of a red-eyed female (WW) with a white-eyed male (w). The F1 males have normal, red eyes (the male gets his only x-chromosome from its mother). Half of the F2 males will have white eyes (from Morgan, 1925: Evolution and Genetics; Printed by permission of Princeton University Press.)

If the frequency of a sex-linked allele is the same in males and females, the population will reach equilibrium in one generation – as in an autosomal gene. If the frequency of a sex-linked allele is not the same in males and females, the equilibrium will not be achieved in one generation (Box 3.4).

The initial difference between the sexes will be halved every generation – with a change in sign, and approach zero in a zig-zag pattern (Fig. 3.7). However, if males and females are pooled, the population as a whole remains at equilibrium.

Since females have two **x** chromosomes and the males have one, the allele frequency at equilibrium is given by $(2q_{\text{females}} + q_{\text{males}})/3$ (Box 3.5).

> ### Box 3.4   Equilibrium frequencies in a sex-linked allele
>
> Mating combinations:
>
> | | | Female genotypes: | **AA** | **Aa** | **aa** |
> |---|---|---|---|---|---|
> | | | Frequencies: | D | H | R |
> | Males | A | $D + \frac{1}{2}H$ | $D^2 + 1/2\,DH$ | $DH + \frac{1}{2}H^2$ | $DR + \frac{1}{2}RH$ |
> | | a | $R + \frac{1}{2}H$ | $DR + \frac{1}{2}DH$ | $RH + \frac{1}{2}H^2$ | $R^2 + \frac{1}{2}RH$ |
>
> Offspring frequencies (for example) $(AA) = D^2 + \frac{1}{2}(2DH) + \frac{1}{4}H^2 = (D + \frac{1}{2}H)^2 = p^2$
>
> If the initial frequency is not the same in the two sexes, the frequency of the allele in generation **n** in males will be equal to its frequency in the females in the previous generation,
>
> $$q_{males,\,n} = q_{females,\,n-1}$$
>
> while its frequency in the females will be equal to the mean of the two sexes:
>
> $$q_{females,\,n} = (q_{males,\,n-1} + q_{females,\,n-1})/2$$
>
> let the difference at any generation be $d_n = q_{females} - q_{males}$
>
> $$\text{then } d_n = -d_{n-1} \text{ or } d_n = (-\tfrac{1}{2})^n d_0$$

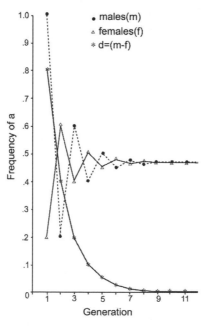

Fig. 3.7 Zig-zag approach to equilibrium in a sex-linked gene when the initial frequency is not the same in males and females. In this extreme example, $q_{males} = 1.0$, $q_{females} = 0.2$. The (absolute) difference **d** is halved every generation. The final equilibrium frequency in this example is (1.0 + 0.4)/3 = 0.47.

Box 3.5   Inheritance of a sex-linked mutation in populations of flour beetles (*Tribolium castaneum*). The marker allele was the recessive allele, paddle (*pd*).

A.   Parent frequencies: $q_{males} = 0.2$, $q_{females} = 0.8$. $d_0 = 0.6$

Offspring:	mutant	normal		
Female	69	172	q^2_{female}	0.286
Male	186	78	q_{male}	0.705
			q_{female}	0.536

$$d_1 = 0.535 - 0.705 = -0.170$$

B.   parent frequencies: $q_{males} = 0.8$, $q_{females} = 0.2$. $d_0 = -0.6$

Offspring:	mutant	normal		
Females	52	176	$q^2_{females}$	0.228
Males	50	181	q_{males}	0.216
			$q_{females}$	0.477

$$d_1 = 0.477 - 0.216 = +0.261$$

In the large and widespread order of insects, Hymenoptera – including bees, wasps and ants – the entire genome is sex-linked: males are haploid (hatch from unfertilized eggs) – and females are diploid. Any mutation will be expressed in the males, and be exposed to natural selection, but will be protected in the heterozygous state in the females. Some writers attach great importance to this haplo-diploid system as a mechanism which allows removal of deleterious mutations quickly.

Literature Cited

Morgan, T.H. 1925. Evolution and Genetics. Princeton University Press, Philadelphia, USA.

Sokal, R.R. and F.J. Sonleitner. 1968. The ecology of selection in hybrid populations of *Tribolium castaneum*. Ecological Monographs 38: 345-379.

Sokal, R.R. and F.J. Rohlf. 1995. Biometry. 3rd edition. Freeman & Co., New York, USA.

Spiess, E.B. 1977. Genes in Populations. Wiley, New York, USA.

Wool, D. and S. Mendlinger. 1973. The *eu* mutant of the flour beetle, *Tribolium castaneum*: Environmental and genetic effects on penetrance. Genetica 44: 496-504.

Wool, D. 1970. Deviations of zygotic frequencies from expectation in eggs of *Tribolium castaneum*. Genetics 66: 115-132.

CHAPTER 4

Deviation from Equilibrium: Genetic Drift – Random Changes in Small Populations

One of the requirements for a genetic equilibrium is that the population be infinitely large. Infinite populations exist only as theoretical concepts, not in nature. How will violating this requirement affect the expected result? Can we predict the genetic fate of a small population in nature?

'EFFECTIVE' POPULATION SIZE

A distinction is often made between real population size, which can be observed and potentially counted, and the 'effective' size as reflected in the number of individuals actually breeding and producing the next generation. Real population size may include individuals above or below reproductive age, which consume some of the resources but do not contribute to population growth and evolutionary change. In practice, the size of natural populations is rarely if ever known or reliably estimated. Another factor that affects the reproductive output is the balance between the numbers of males and females. The number of females sets a limit on population growth of sexually-reproducing organisms: a male may mate with more than one female, but an excess of males may not contribute to the next generation as much as may be implied by their number. The effective size N_e is defined as

$$N_e = 4N_mN_f/(N_m + N_f)$$

N_e is smaller than N. Some numerical examples are given in Table 4.1. The table shows that the real population size may not reflect the breeding potential of the population when the sex ratio departs from 1:1.

Table 4.1 Effective population sizes when the sex ratio is different from 1. Real population size is taken as 100

Females	males	N_e
50	50	100
75	25	75
90	10	36
95	5	9
98	2	8
99	1	4

SAMPLING ERRORS

As N becomes smaller, the effects of chance on allele frequencies become more pronounced. For example, imagine a population of N = 10 individuals, composed of 5 **AA**, 4 **Aa**, and one **aa**. An accidental death of the **aa** individual will change the allele frequency q_a from 0.33 to 0.22, and genotype frequencies at equilibrium will consequently change from 0.49: 0.42:0.09 to 0.61:0.34:0.05. Accidental deaths (i.e., unrelated to genotype) may perhaps be the greatest source of mortality in natural populations of any size, but they must be important in particular in small populations. Readers with some familiarity with probability theory will recall that in a gene pool of which the allele frequency of **A** is **p**, it is possible to calculate the probability of sampling at random N pairs of gametes (forming N diploid zygotes) of which 0, 1, 2...N–1, N will be **AA**, from the binomial distribution. If **p** = **q** = 0.5, only 68% of the resulting sample will lie in the range of **p** = 0.45 to 0.55, and 32% will be farther away from the true mean of 0.5. Sampling variation can be an important source of deviation from expected Hardy-Weinberg proportions when the population is small.

GENETIC DRIFT IN SMALL POPULATIONS

The American geneticist and evolutionary biologist, Sewall Wright, was the first to call attention to the effects of chance on genetic composition in small populations, in a series of papers published in the 1930-1940s. These effects were first considered negligible (and referred to as the 'Sewall Wright effect'), but later recognized as potentially very important. These random effects – called **genetic drift** and **random fixation** – are now considered a most important driving force in evolution, at least at the molecular (DNA) level.

It is possible to simulate these effects by the use of a simple, familiar toy, a slanted board into which nails are driven. A marble is released at the top of the board, rolls down and hits the nails in its path, changing course and velocity at random, until it reaches the bottom. At the two sides of the board are grooves, which trap the marble if it reaches either side.

In this model, the width of the board represents the allele frequency **q** of the marker, from zero to one, and the length of the board, from top to bottom represents time (Fig. 4.1).

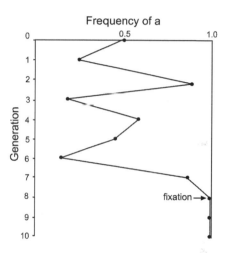

Fig. 4.1 Simulation of random genetic drift in a small population. The position of the points in the erratic path relative to the width of the board represents the marker allele frequency in the population at sequential points in time. Allele frequencies change at random until they reach fixation at one of the sides.

The position of the marble relative to the width of the board represents the allele frequency at any moment in time. As any trial will show, the path of the marble is erratic and depends on many variable factors, like its direction and velocity at the time it strikes each nail (although in theory these factors can be measured, in practice the marble path is unpredictable). This erratic path simulates **genetic drift**. The marble drifts back and forth on the board and must eventually reach either side (in genetic terms, it reaches either **q** = 0 or **q** = 1). Once in the groove, the marble cannot get back on the board. In genetic terms, once it has lost one of the two alleles, the population is no longer variable (until a new mutation occurs). This is a case of **random fixation**. The time of fixation can be short or long, but fixation must eventually occur.

Wallace (1969) suggested a simple game to illustrate genetic drift and random fixation (I used it successfully in my courses). A series of 9 boxes is set up, containing large numbers of marbles of two colors, say white and blue, in proportions 0.1, 0.2, 0.3 … 0.9 of blue. One then decides on a small sample size, say 10. Starting with the box with **q** = 0.5, a random sample of 10 is drawn from the box and the number of blue is recorded. The sample is then returned to the box and mixed. The next sample is taken from the box corresponding to the <u>proportion of blues in the previous sample</u>. The process is continued – each sampling event representing a generation – until the sample contains marbles of only one color (white or blue): the population has reached fixation.

This simple game illustrates that the genetic fate of the population is determined at each generation by the (chance) composition of the gamete pool, represented by the sample – regardless of how variable was the parent population. It can be used to illustrate that fixation rate depends on sample size.

Sewall Wright showed that random fixation affects the shape of the allele frequency distribution. Starting with a large number of small populations of size N at time 0, with a mean of **q** = 0.5, the distribution of allele frequencies is initially unimodal and approximately normal, but if the same populations are followed in subsequent generations, more and more of the populations go to fixation and the distribution becomes U-shaped (Fig. 4.2). When time goes to infinity, all populations should be fixed.

Wright showed that the rate of fixation depends on sample size. 1/2N of small populations of size N are expected to become fixed every generation. When N is large, the rate of fixation becomes small, and may not be noticeable in practice.

GENETIC VARIATION IN SUBDIVIDED POPULATIONS

A population may seem large, but be subtly subdivided into small units. This may have important effects on its genetic structure. The genetic variance in a subdivided population can be partitioned into an among-subunits and within-subunits component. The variation among subunits is expected to increase with time. At time **t** after the founding of the population, the variance is expected to be

$$\sigma^2 = \mathbf{pq}\,(1 - (1 - 1/2N_e)^t$$

where **p** and **q** are the average allele frequencies in the subdivided population, and N_e the average subunit population size (the expected genetic variance in an infinite population is $\sigma^2 = \mathbf{pq}$ and will not change with time). A useful measure of the change in variance is the 'standardized

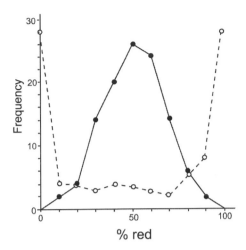

Fig. 4.2 Allele frequency distribution of small (N = 10) populations started at a common frequency **p** = **q** = 0.5 and affected by random processes in a laboratory demonstration of Wallace's experiment. After one generation of random drift, the distribution is approximately normal (solid line). After 10 generations the distribution tends to become U-shaped as more and more of the population are fixed at **q** = 0 or 1.

variance' $f = s^2/pq$, where s^2 is the observed variance among subunits (Crow and Kimura, 1970).

To illustrate this effect, experimental populations of flour beetles were subdivided into 15 subunits, each containing 1/15 of the food supply and colonized by 1/15 the number of beetles as the control, undivided population: four pairs in each subunit, 60 pairs in the control (Wool and Mendlinger, 1981). Mixed populations were set up with equal numbers of wild-type and black beetles (two pairs of each per subunit). The allele frequencies at the black (**b**) marker locus in genetically 'mixed' populations were monitored (in addition to a number of ecological variables, which were also measured in genetically-homogeneous populations of the same structure). As expected, the standardized variance **f** increased in the subdivided populations (Fig. 4.3).

THE WAHLUND EFFECT

In a subdivided population, the <u>mean</u> genotype frequencies (of all subpopulations) will deviate from Hardy-Weinberg expectation even if the population within each subunit is at equilibrium.

Let the genotype frequencies in subunit i be at equilibrium (p_i^2, $2p_iq_i$, q_i^2). The mean allele frequencies will be $p = \Sigma p_i/n$ and $q = \Sigma q_i/n$ for

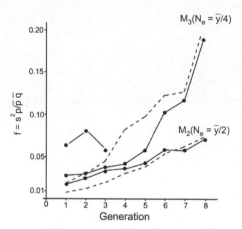

Fig. 4.3 Increase of the standardized variance $f = s^2_p/pq$ in subdivided populations of flour beetles (solid lines). The broken lines are the expected values of f if the effective population sizes N_e were ½ or ¼ of the observed N (after Wool and Mendlinger, 1981).

n subunits. If we then calculate the mean <u>genotype</u> frequencies, we will expect to find the following relations,

$$\text{Frequency of } \mathbf{AA} = \Sigma p_i^2/n = \mathbf{p}^2 + \sigma^2_p$$
$$\text{Frequency of } \mathbf{Aa} = \Sigma(2p_iq_i)/n = 2\mathbf{pq} - 2\sigma^2_p$$
$$\text{Frequency of } \mathbf{aa} = \Sigma q^2_i/n = \mathbf{q}^2 + \sigma^2_p$$

(σ^2 is the variance of allele frequencies among subunits). In the population as a whole, there will be a deficiency of heterozygotes and an excess of homozygotes compared with expectation – although each subunit separately is at equilibrium. The magnitude of these deviations will be proportional to the genetic variance among subunits. This was shown in 1928 by Wahlund (Li, 1956).

This result is important, since in Nature we are seldom – if ever – aware of the intra-structure of a population. Deficiencies of heterozygotes and excess of homozygotes may be misinterpreted as the result of other processes, like inbreeding or selection, while in fact they may be due to subdivision.

THE FOUNDER PRINCIPLE

Random processes may have a profound effect on the evolutionary fate of a new, colonizing population, because it depends on the gene pool of the (usually small) sample of founding individuals. This effect is termed 'the founder principle' (Mayr, 1963).

A population may be started at a new site by a small number of founders – perhaps as small as a single pair or a single fertilized female. Propagule size may have a decisive effect on the future composition of the population, both genetically and ecologically. Imagine an 'empty niche' somewhere – like an island, which has not been colonized but where a population can be sustained. This suitable site may be located in a large, unsuitable area for that organism – an ocean for terrestrial organisms, a mountain top, or an oasis in a desert. Then imagine, at a distance from that island, a source (continent) where a large and genetically-variable population of these organisms thrives. The distance, and the unsuitability of the surrounding habitat, necessarily limits the numbers of colonizers that arrive at the island, and these immigrants then face other difficulties before establishing a colony. Colonization of Daphne, an island in the Galapagos archipelago, by one species of 'Darwin's finches' was followed by Grant and Grant (1995) for more than 25 years. Small numbers of birds were seen on Daphne from time to time – arriving probably from the nearest neighboring island, Santa Cruz, 8 km away – but for 22 years no nesting pairs were established. In 1983, the first nests were seen, and the numbers of breeding pairs subsequently increased slowly from 3 to 22 – some of them new immigrants, and some the offspring of resident pairs (Grant and Grant, 1995).

The small group of founders is likely to be less variable genetically than the parent population. Even if all genotypes have the same (small) probability of reaching the island and reproducing there, the sample actually arriving is likely to be different from the source because genotypes may be lost by chance during the voyage. Due to its small size, the colonizing group cannot contain all the genotypes that existed in the source population. The ability of the population to reproduce, grow, and compete with the existing populations of other species, may depend on the chance mixture of genotypes among the founders. The founders may in turn affect the environment at the site, making it ecologically difficult for future immigrants – whether conspecific or heterospecific – to get a foothold and reproduce there.

The founder principle is considered an important factor in evolution, although situations like those reported by Grant and Grant (1995) are rarely recorded. Evidence that the numbers of founders were small can be gathered a posteriori from studying the genetic structure of island populations. The snail *Theba pisana* was studied on a small island west of Australia (Johnson, 1988). This snail is of Mediterranean origin, most probably transported to Australia involuntarily by man. A survey of protein variation of the snail at several locations around the island, and in samples from Mediterranean and European countries (France, England and Israel) which could have been the source of the founders of the island populations, indicated a lot of variation among sites on the island. The mobility of the snails being limited, the results are in accordance with the assumption of several colonization events. The data suggest that the source of the snails may have been in France (Johnson, 1988).

In a recent study, the colonization of South and North America by *Drosophila subobscura*, which was considered a Mediterranean species until recently, was investigated. This species was discovered in South America in 1978 (Mestres et al., 2004). The populations of *D. subobscura* from the New World were studied using two markers – a chromosomal inversion O_5, which contains a lethal gene, and the Octanol dehydrogenase gene (ODH) also located in (or near) the inversion. The comparison of the sequenced ODH gene from many populations in North America led to the conclusion that all the studied O_5 inversions carried copies of the same haplotype of ODH and the same lethal gene. The South American samples showed similar associations. The authors concluded that a small founding population of *D. subobscura* (estimated at 10-15 individuals) which carried the O_5 inversion as well as the lethal gene, arrived at either South or North America, and multiplied there. A larger group later spread from there to reach the other continent (Mestres et al., 2004).

PROPAGULE SIZE OR GENE POOL SIZE?

A small number of founders necessarily carry a limited gene pool. However, which is the more important determinant of the fate of the future population – the numerical (ecological) presence of a small number of individuals, or the long-term genetical effect of a small gene pool?

A recent attempt to study the effect of propagule size on colonization success in a natural environment was made by Memmott et al. (2005) in New Zealand. Fifty-five propagules of size 2, 4, 10, 30 and 270 individuals of the psylla, *Arytainilla spartiophila*, imported from England, were released along a 135-km transect in patches of the weed, *Cytisus scoparius* (broom) as part of a weed control program. The release sites were monitored for six years. There was a positive correlation between the size of the propagule and the probability of colonization success in the first year, but not with the probability of establishing a permanent colony: sites where the psylla survived the first year, had a 96% probability of long-term colonization success regardless of the original propagule size (even a single mated psylla was able to found a colony due to its high reproductive potential). Of course, for the colonizing insects, this study was very favorable because they were deliberately released in sites where the food plant was growing vigorously.

To the best of my knowledge, Dawson (1970) performed the only experiment specifically designed to answer the question of whether propagule size or gene-pool size are the more important. He first created source populations of *Tribolium castaneum* founded by 1, 2, 4, 8, 16, 32 and 64 pairs of beetles and let them reproduce. These source populations contained gene pools of different sizes. From each source, he then created his experimental replicates – each of which was founded by 10 pairs. To each replicate he added 10 pairs of a competitor, *T. confusum*. The variable of interest was the time it took until the competitor was eliminated.

The result clearly demonstrated that populations founded with a small gene pool were less successful in competition (Fig. 4.4) (the propagule size was the same in all replicates).

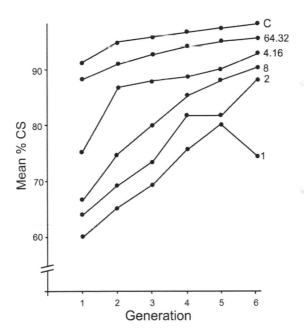

Fig. 4.4 The effect of the founder gene pool size on subsequent competitive success of flour beetles. Each line was founded by ten pairs of *Tribolium castaneum* (CS), competing with 10 pairs of *T. confusum* (CF). The CS founders came from source populations initiated with 1, 2, 4...64 pairs of parents (as marked in the figure). The ordinate is the percentage of CS in the adult output of every generation. The larger the gene pool, the faster was the rate of increase of CS (after Dawson, 1970).

Literature Cited

Crow, J.F. and M. Kimura. 1970. An Introduction to Population Genetics Theory. Harper and Row, New York, USA.

Dawson, P.S. 1970. A further assessment of the role of founder effects in the outcome of *Tribolium* competition experiments. Proceedings of the National Academy of Science, USA 66: 1112-1118.

Grant, P.R. and B.R. Grant. 1995. The founding of a new population of finches. Evolution 49: 229-240.

Johnson, M.S. 1988. Founder effects and geographic variation in the land snail *Theba pisana*. Heredity 61: 133-142.

Li, C.C. 1956. Population Genetics. Chicago University Press, Chicago, USA.

Mayr, E. 1963. Animal Species and Evolution. Harvard University Press, Boston, USA.

Memmott, J., P.G. Craze, H.M. Harman, P. Syrett and S.V. Fowler. 2005. The effect of propagule size on the invasion of an alien insect. Journal of Animal Ecology 74: 50-62.

Mestres, F., L. Abad, B. Sabater-Munoz, A. Latorre and L. Serra. 2004. Colonization of America by *Drosophila subobscura*: association between ODH gene haplotypes, lethal genes and chromosomal arrangements. Genes Genetic Systems 79: 233-244.

Wallace, B. 1969. Topics in Population Genetics. Norton, New York, USA.

Wool, D. and S. Mendlinger. 1981. Ecological and genetical consequences of subdivision in *Tribolium* populations. Journal of Animal Ecology 50: 421-433.

CHAPTER 5

Deviations from Equilibrium: Mutations

HISTORICAL REVIEW

Mutations are changes in the genetic material (DNA). In the past, mutations were recognizable only when they were expressed in the phenotype of an individual. Today, with the current methods of DNA sequencing and sequence comparisons, the occurrence of mutations can be documented whether or not they are expressed in the phenotype.

That heritable changes occurred ('spontaneously') was already well known to breeders of livestock in the 18th and 19th centuries (Wood and Orel, 2001). The individuals expressing a changed phenotype (called 'sports') were used by animal and plant breeders selecting for desirable characters (Darwin, 1868). However, the nature and causes of these changes – which today are referred to as mutations – remained unknown.

With the rediscovery of Gregor Mendel's paper on the hybridization of garden peas in 1900, and the subsequent growth of genetics as a science, the interest in mutations rose greatly, together with the understanding of their causes. The Dutch botanist, H. de Vries, and his supporters advocated the idea that new species appeared in evolution suddenly, via mutations, contrary to the strict Darwinian doctrine of gradual evolution by natural selection (see in later chapters). Ways were discovered to induce mutations in laboratory populations of *Drosophila* (by T.H. Morgan and his colleagues in America). But it was not until the discovery of the structure of DNA in the 1950s by Watson and Crick (Watson, 1969) that mutations began to be understood.

ABOUT MUTATIONS: TERMINOLOGY

The reader will recall that the DNA molecule is a double-helical chain, composed usually of four units (nucleotides): Adenine (A), Guanine (G), Cytosine (C) and Thymine (T). The genetic information is coded in the chain by the <u>order</u> in which the nucleotides are arranged, and 'read' as sequential groups of 3 nucleotides (**triplets**). When the DNA is translated to produce proteins, each triplet (**codon**) codes for a specific amino acid. These are joined together to produce a peptide. Changes in the DNA, which result in replacement of one nucleotide by another, are called **point mutations**. Point mutations can be **synonymous** if the changed triplet still codes for the same amino acid, due to the 'degeneracy' of the genetic code (Table 5.1). If the changed codon causes the replacement of one amino acid by another, it is referred to as a **substitution** mutation. Some mutations involve more than one nucleotide. These are **duplications**, **deletions** or **insertions** of a segment of DNA (Fig. 5.1).

Table 5.1   The 'universal' genetic code: the 64 codons in mRNA, and the amino-acids they code for (in italics)

code	acid	code	acid	code	acid	code	acid
UUU, UUC	phe	UCU, UCC	ser	UAU, UAC	tyr	UGU, UGC	cys
UUA, UUG	leu	UCA, UCG	ser	UAA, UAG	stop	UGA, UGG	Stop trp
CUU, CUC	leu	CCU, CCC	pro	CAU, CAC	his	CGU, CGC	arg
CUA, CUG	leu	CCA, CCG	pro	CAA, CAG	gln	CGA, CGG	arg
AUU, AUC, AUA	ileu	ACU, ACC	thr	AAU, AAC	asn	AGU, AGC	ser
AUG	met	ACA, ACG	thr	AAA, AAG	lys	AGA, AGG	arg
GUU, GUC	val	GCU, GCC	ala	GAU, GAC	asp	GGU,GGC	gly
GUA, GUG	val	GCA, GCG	ala	GAA, GAG	glu	GGA, GGG	gly

a.	…. CTTGCAACGTAGAT….
b.	…. CTTACAACGTAGAT….
c.	…. CTTACAACG<u>ACG</u>TAGAT….
d.	…. CTTACAACGA…..GAT….

^CGTA^

Fig. 5.1   Types of mutations (after Li and Graur, 1991). a. Original sequence. b. Point mutation. c. Duplication. d. Deletion.

The modern methods of molecular genetics enable the detection of all types of mutations when two or more sequences of DNA are aligned and compared. Mutations are detected as the difference between sequences of the same section of DNA in ancestral versus extant DNA (e.g., from different organisms when the phylogeny is known), or between two extant sequences if the ancestral sequence is unknown (Graur and Li, 2000). Understandably, mutations causing genetic disorders in humans attract the most attention. For example, some of the earliest to be investigated were point mutations in the α- and β- globin-coding genes, components of the hemoglobin molecule, which are the causes of different kinds of anemia. Considerable expense and many years of research resulted in the recent completion of the ambitious Human Genome Project, which now enables the mapping of known genetic disorders.

THE FREQUENCY OF MUTATIONS

Estimates of mutation frequency run between 10^{-4} and 10^{-10} per site per generation. Despite being quite rare, mutations affect long-term evolutionary processes profoundly. By recombination during the formation of the gametes (sexual cells) and fertilization, the rare mutations are reshuffled and enable the formation of many more genotypes, which constitute the material for natural selection. (It should be emphasized that mutations are only relevant to evolution if they occur in the sexual cells and can be transmitted to the next generation. Mutations that occur in somatic cells may change the phenotype of the individual, but have no evolutionary consequences.) In cases of asexual reproduction, new deleterious mutants may be protected (as heterozygotes) against removal by selection and accidental extinction, and mutations may thus accumulate. The fast spread of drug resistance in disease-causing bacteria and viruses is due to their asexual reproduction and short generation time, allowing rare mutants surviving drug application to multiply and spread.

When molecular evolution became the dominant branch of modern evolutionary research, at the close of the 20th century, mutations became the focus of intensive theoretical research (Graur and Li, 2000). The main driving force in molecular evolution is thought to be random fixation of 'neutral' or slightly-deleterious mutations (see Chapter 18).

However, it should be emphasized that for short-term processes, like laboratory experiments lasting a few generations, or observations on a natural population lasting a few years, the chance that deviations from Hardy-Weinberg equilibrium are caused by mutation is negligible.

THE FATE OF A SINGLE MUTATION

The fact that mutations are rare makes it likely that most of them will not occur more than once: the probability that a mutation which occurs in 10^{-4} of the gametes will occur twice is 10^{-8}. What is the chance that a single mutation, occurring once only, will be lost – or will survive and spread?

The calculations in Box 5.1 show that a single 'neutral' mutation has a 37% chance of being lost in the first generation. In two generations, this probability increases to 53%. Of course, if the mutation is deleterious, the chance of loss is even greater.

Box 5.1   **Calculation of the probability that a single mutation be lost**

In a population of 2N gametes, one mutation from **A** to **a** will produce a single **Aa** individual. If that individual reaches sexual maturity, it will have to mate with an **AA** individual. If the pair produces no offspring, the mutation will of course be lost. If a single offspring is produced, there is a 50% probability that the mutation will be lost since half of the offspring of this mating will be **AA**. If there are 2 progeny, the probability of loss is 25%. In general, for y progeny,

$$p_{loss} = (\tfrac{1}{2})^y = 1/2^y$$

Let p_y be the probability that a pair leaves y offspring. The probability of loss then becomes

$$p_{loss} = \Sigma p_y * (\tfrac{1}{2})^y$$

As a reasonable approximation, we can assume that the probability of leaving y offspring (y = 0, 1, 2, 3...) is random. From the Poisson distribution (consult a book on statistics, e.g. Sokal and Rohlf, 1995), we get

$$p_y = Y^y/e^{Y} * y!$$

where Y is the average number of offspring per pair, e the base of the natural logarithms (e ~ 2.7183...), and y-factorial (y!) is the product 1*2*3*...y).

In nature, population size is considered stable over large periods of time. This implies that on average, each pair is replaced by one pair of offspring (Y = 2). Inserting Y = 2, we get

$$p_y = 2^y/e^2\, y!$$

and $$p_{loss} = (2^y/ e^2\, y!) * 1/ 2^y = 1/e^2 * \Sigma\, 1/y!$$

The expression $\Sigma 1/y!$ adds up to $(1 + 1/2! + 1/3! + 1/4!...) = e$

so that $p_{loss} = 1/e = 0.3679...$

RECURRENT MUTATIONS

If the mutation recurs, it will not be lost and its frequency may theoretically increase, although the process may be very slow and require thousands of generations, depending on the mutation rate. If the mutation rate is μ from **A** to **a**, the allele frequency in generation n+1 is

$$q_{n+1} = \mu(1 - q_n) + q_n$$

If recurrent mutations occur in both directions at different rates – a balanced equilibrium may theoretically be reached in which the allele frequency depends only on the mutation rates in the opposing directions (consult Li, 1956 for more details).

A balanced mutational pressure was recently suggested to best explain the stable, unusually low GC content – in the base composition of DNA of the aphid symbiotic bacterium, *Buchnera aphidicola* (Wernegreen and Funk, 2004). The authors discovered that the rate of GC $\rightarrow$ AT and AT $\rightarrow$ GC synonymous substitutions in the bacterial DNA were equal, and there was no evidence for selection.

GENETIC LOAD

It is accepted that most of the mutations in natural populations are slightly deleterious or neutral (do not affect the fitness of their carriers. See Chapter 18). Most of the mutations that people were aware of before the 1960s – in particular in humans and *Drosophila* – had strong phenotypic effects and many were congenital defects. Mutations were thus considered mostly harmful. This has led people to conclude that natural selection should have removed inferior genotypes during thousands of generations of evolution, and that only those organisms where all the internal physiological systems and organs work in harmony survived to date. The chance that any new change will be favorable was considered small, because a change is likely to disrupt the harmony and cause damage. Recessive alleles that remain in the genome in heterozygous condition may be considered a negative load ('genetic load') on the population, with the potential to cause harm when they occasionally appear in the homozygous condition by recombination.

Research by Dobzhansky and his colleagues in natural *Drosophila* populations revealed that the frequencies of deleterious alleles may be as high as 40% of the tested chromosomes (Dobzhansky, 1971; Lewontin 1974). Dobzhansky summarized his impression in these words:

A man or a fly free of genetic loads might perhaps be a superman or a superfly, but as far as anybody knows, such a prodigy never walked the earth (Dobzhansky, 1961, p. 296).

Literature Cited

Darwin, C. 1898 (1868). Variation in Animals and Plants under Domestication. Appleton & Co., New York, USA.

Dobzhansky, T. 1961. Mankind Evolving. Yale University Press, Boston, USA.

Dobzhansky, T. 1971. Genetics of the Evolutionary Process. Columbia University Press, New York, USA.

Graur, D. and W-H. Li. 2000. Fundamentals of Molecular Evolution. Sinauer, Sunderland, USA.

Lewontin, R.C. 1974. The Genetic Basis of Evolutionary Change. Columbia University Press, New York, USA.

Li, C.C. 1956. Population Genetics. Chicago University Press, Chicago, USA.

Li, W-H. and D. Graur. 1991. Fundamentals of Molecular Evolution. Sinauer, Sunderland, USA.

Sokal, R.R. and F.J. Rohlf. 1995. Biometry. 3rd ed. Freeman, New York, USA.

Watson, J.D. 1969. The Double Helix. The New American Library, New York, USA.

Wernegreen, J.J. and D.J. Funk. 2004. Mutation exposed: a neutral explanation for extreme base composition of as endosymbiont genome. Journal of Molecular Evolution 59: 849-850.

Wood, R.J. and V. Orel. 2001. Genetic Prehistory in Selective Breeding: a prelude to Mendel. Oxford University Press, Oxford, UK.

CHAPTER 6

Deviations from Equilibrium: Migration

Migration is the movement of individuals from site A to site B and/or back. It is a directional change, while **dispersal** is movement from point A in all directions. In the literature, this distinction is not always made.

MIGRATION AS AN ECOLOGICAL PHENOMENON

Ecologically, migration and dispersal mean the transport of **propagules** (small units which can reproduce and establish a population) from place to place. Propagules may be as small as pollen grains, fruits, seeds, dormant eggs or any other unit that may give rise to a new organism away from the starting point – and of course may also be one or more whole organisms.

In human history as well as today, migration played a major role – in colonizing new lands (America and Australia are relatively recent examples), and in reshuffling and mixing human populations. Many animal species migrated to new lands within recent historical times. Some were brought over voluntarily by human immigrants: cattle, goats, pigs, cats and dogs, were introduced to oceanic islands. The starling and domestic sparrow were brought with the Europeans to America and Australia "to remind them of their homeland", and became widespread in the new lands. A famous case is the spread of the European rabbit (*Oryctolagus cuniculus*) in Australia: from a small colony (two dozen) brought in 1859 from England, populations increased to hundreds of millions in 20 years or so, consumed and destroyed grasslands, and threatened the local sheep-wool industry. The epidemic was stopped by introducing a myxomatosis virus from Brazil. The virus is harmless to the native rabbits in Brazil but kills European rabbits.

Some unwanted species, like rats, flies and other pests accompanied migrating humans. The process goes on today, with exotic birds escaping from aviaries and taking over native habitats, causing concern for conservationists. With the development of human means of transport, like ships and airplanes, untold numbers of organisms are spreading all over the world. Locally, immigration often takes the form of colonizing sites that have been deserted or recently exposed. Some good examples are the colonization of land reclaimed from the sea in the Netherlands by plants, and the recolonization of the island of Krakatoa, which was destroyed by volcanic eruption in 1883 and a 'new' island emerged in its place (Thornton and Rosengren, 1988). In many countries, getting rid of 'exotic' species – or minimizing their impact – is a major concern of conservationists.

Man is using immigration as a tool in the fight against pests of agriculture. Controlled immigration of parasites and predators is a widespread method of biological control. In Israel there have been several major successes, in particular against pests of citrus orchards. The scale *Pseudococcus citriculus* was almost eliminated from the orchards by the introduction of a small parasitic wasp, *Clausenia* sp., from Japan. The black and red scales are similarly controlled in Israel and the USA by minute wasps of the genus *Aphytis*, imported from Hong Kong.

Annual migration of birds is a well-known phenomenon since biblical times. The spectacular sight of the annual migration of large herds of African mammals has become publicized by television programs and must be familiar to viewers all over the world. Similar migrations must have taken place in the remote past as well. Early 19th century geologists in Europe were struck by the appearance of bones of elephants, rhinoceroses, hippotami and other tropical animals in rock formations in Europe. Georges Cuvier regarded the fossils as evidence of catastrophic changes in climate in the history of the world. The British geologist Charles Lyell suggested instead that large mammals could have reached Europe by migration:

> "the geologist…may freely speculate on the time when herds of hippopotami issued from North African rivers, such as the Nile, and swam northwards along the coast of the Mediterranean… here and there they may have tarried awhile where they landed, to graze or browse, and afterwards have continued their course northwards…". Lyell, 1873, p. 209).

GENETIC AND EVOLUTIONARY CONSEQUENCES OF MIGRATION

Genetically, migration is the exchange of genes between populations – a phenomenon referred to as **gene flow.** In contrast with the laboratory, natural populations do not normally have borders that restrict gene flow. The extent of gene flow depends on the dispersal ability of the organisms which carry their genes between populations. If the populations are not genetically identical – an unlikely situation – migration can cause a genetic change in the recipient population, the source population, or in both.

In human history, migration has been a major cause of genetic changes in many countries, via movements of armed forces during wars, occupation, transfers of slaves and prisoners-of-war or marriages. Countries like the USA and Israel are composed of migrants from many different populations, and the influx of immigrants (and migrant genes) to Europe, USA and Australia from other parts of the world continues.

THE FATE OF A SINGLE IMMIGRANT

In nature, it is difficult to monitor the genetic contribution of an immigrant to the gene pool of the recipient population because its progeny cannot be identified separately from other progenies. This difficulty can be overcome in carefully planned laboratory experiments.

There are some important differences between the effects of an immigrant and that of a mutation on the recipient population. A mutation changes the allele frequencies at a single locus in a single individual. By contrast, a single immigrant carries with it an entire genome, and can potentially incorporate many new alleles in the gene pool of the recipient population. Also, a single immigrant – particularly if it is a male – can sire many progenies and make a much larger impact on the structure of the recipient population than a single mutation event can. Finally, migration is a far more frequent event than mutation.

EXPERIMENTAL EVIDENCE

Merrell (1965) established *Drosophila* populations – each with 200–400 individuals – which carried different, <u>dominant</u> genetic markers. Into each population he introduced a single migrant wild type (++) male (note that the + allele was recessive to the markers). The initial frequency of + was no greater than 0.005. But in all populations, the frequency of the markers decreased as more ++ progeny were produced, and the wild type eventually dominated all populations. The carriers of the mutant markers seemed unable to compete with the wild type flies just as domestic animals – albino rats, mice, or rabbits for example – survive only when protected by man, and will not succeed if released into nature.

Sokal et al. (1974) prepared 40 populations of the flour beetle, *Tribolium castaneum*, each initiated with 500 wild-type individuals. Among them was one female, which was previously mated to a homozygous black (**bb**) mutant male. The initial frequency of **b** in the populations was therefore 0.001. Unlike the (expected) fate of a single mutant (Chapter 5), the **b** allele did not become extinct, but prevailed at a low frequency in 26 of the 40 populations for the duration of the study.

Kaufman and Wool (1992) used homozygous recessive mutant populations of the flour beetle *Tribolium castaneum* as recipients, to monitor the fate of immigrants. The genetic marker was *pearl* (**p**), a mutation that eliminates the dark pigment in the central ommatidia in the compound eye, so that the homozygous mutant has white eyes rimmed with a black frame. Single immigrants (later also larger numbers) of four different wild-type strains were introduced into replicate populations of *pearl* and allowed to mate for one week – afterwards all adult beetles were removed and the eggs they laid were allowed to develop into adults. Genetically, all the offspring of the immigrants were heterozygotes +**p** and therefore had wild type, dark eyes, easily separated from the offspring of resident individuals, which were homozygous **pp** and had white eyes.

The immigrants invariably left more offspring than expected from their frequency in the parent population. Kaufman and Wool argue that the advantage of the immigrants could be due to assortative mating – with immigrants being preferred as mates, either because they carried favorable wild-type characters or because they were rare; but could also be due to hybrid vigor: offspring of the immigrants were necessarily hybrids, and thus were heterozygous in loci which may have been deleterious in the **pp** homozygotes.

Note: The results of these experiments may create the impression that single migrants have a good chance of successfully contributing to the gene pool of the recipient population in nature. This impression may not be justified. The experiments reported here consider only those migrants that did arrive at the target population: in nature many migration attempts fail. Further, the experiments ignore the ecological differences between the habitats of the source and the new site: the migrants and the resident *Tribolium* beetles were reared for many generations in exactly the same standard food medium. This is unlikely to be the case in nature. The migrants can be expected to have a lower fitness in the new environment than the residents. Via et al. (2000) showed that when transferred between alfalfa and clover, translocated strains of the pea aphid suffered extremely low fitness compared with the resident strains on the two host plants.

THEORY: UNIDIRECTIONAL RECURRENT MIGRATION

If immigrants repeatedly arrive into a recipient population, from a large source with different allele frequencies, immigration will eventually equalize the frequencies with the source population (Fig. 6.1).

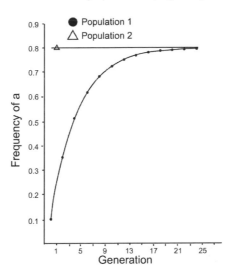

Fig. 6.1 Unidirectional recurrent migration from a source to a recipient population. In this example, $m = 0.2$. Within about 25 generations, the allele frequency in the recipient population will be the same as that of the source.

Let a proportion **m** migrate from population 2 (source) into population 1 (recipient) every generation. Further assume that population 2 is large enough so that its allele frequencies remain stable and are not affected by the emigration. After immigration, population 1 will be composed of **m** migrants and $1 - m$ local residents. Genetically, after migration

$$P_1 = m\ p_2 + (1 - m)\ p_1, \text{ or } P_1 = p_1 + m\ (p_2 - p_1)$$

where **P** represents the frequency at the end of the process and **p** the starting frequencies. The final results thus depend on **m** as well as on the difference between p_1 and p_2. If the final and initial frequencies can be estimated from other data, then **m** could be estimated from the formula

$$m = (P - p_1)/(p_2 - p_1)$$

This formulation was used by Wallace (1969) to calculate the rate of gene flow from the white into the black population of America during the 10 generations of slavery (about 300 years since the first slaves were brought from Africa). He used mainly blood-group allele frequencies as markers. The frequencies of the markers in African populations were taken as p_1, and the frequency in the white American populations as p_2. The current black American population was considered P. Most markers yielded estimates of m close to 3% per generation.

Recurrent migration in opposite directions can lead to a stable equilibrium. The rate of approach to equilibrium depends on the difference in marker frequencies between the recipient and source population and on the proportion of migrants exchanged between them, m_{12} and m_{21} (Box 6.1).

Box 6.1 Equilibrium frequencies with recurrent migration. P_1 and P_2 are the final equilibrium frequencies; m_{12} and m_{21} are the migration rates.

After one generation of migration,

$$P_1 = (1 - m_{12})\, p_1 + m_{21}\, p_2$$
$$P_2 = (1 - m_{21})\, p_2 + m_{12}\, p_1$$

the difference between the two populations will be

$$P_1 - P_2 = (p_1 - p_2) - [2(m_{12}\, p_1 - m_{21}\, p_2)]$$

The difference will diminish every generation (as the formula shows) by the amount enclosed in the square brackets. At equilibrium,

$$m_{12}\, P_1 = m_{21}\, P_2$$

EMIGRATION AND GENETIC CHANGE IN THE SOURCE POPULATION

Narise (1968) experimented with laboratory populations of two strains of *Drosophila*: a wild type and a mutant strain *vestigial (vs)* which has deformed wings and cannot fly. The experimental arena was composed of interconnected jars with food. Narise released 100 flies into the central jar in every replicate, but varied the proportion of *vs* from zero to 100%. When all the flies were *vs*, they remained in the central jar and did not migrate. When all were ++, about 46% of them left and entered the other connected jars. When both strains were introduced in different numerical proportions – keeping the same density – almost always *vs* migrated more than ++ despite its handicap, changing the genotype frequencies in the central jar (Fig. 6.2)

ESTIMATING GENE FLOW IN NATURAL POPULATIONS

Two methods are commonly employed to estimate gene flow in natural populations. In both methods, the criterion is Nm, where N (or Ne) is the (effective) population size, and m is the proportion of migrants of the total population. The methods do not yield estimates of either m or N, but only

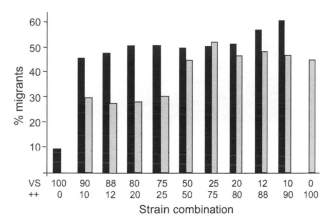

Fig. 6.2 Proportions of wild-type and the flightless *vs* mutants emigrating from the central jar when introduced to it in different ratios of *vs* / ++. Note that more *vs* (black bars) than ++ emigrated in the mixed cultures (after Narise, 1968).

their product **Nm** is an estimate of the mean number of migrants which were exchanged between the populations per generation.

WRIGHT'S F STATISTICS

The older, and more often used, of the two methods is Wright's F-statistics (first published in 1941; Cockerham and Weir, 1993), which are based on the inbreeding coefficients (see the next chapter). If there is no gene flow between subunits in a subdivided population, they will drift genetically apart and will become more and more inbred – and therefore different from each other, especially if selection also operates in different subunits. These processes will increase the variance among subunits. With increasing levels of gene flow, the differences among subunits will be reduced. Wright showed that the variance due to inbreeding in a population can be partitioned into components,

$$F_{it} = F_{is} + F_{st}$$

where F_{it} is the total genetic variance in the population, F_{is} is the variance component among individuals within subunits, and F_{st} is the variance component among subunits. Wright showed that F_{st} is directly related to **Nm**:

$$F_{st} = 1/(4Nm + 1)$$
$$\text{or} \quad Nm = (1 - F_{st})/4F_{st}$$

Further, he showed that if **Nm** is smaller than 1, then the differences among subunits may be caused by genetic drift alone. If **Nm** >1, then factors other than drift – especially selection – can be suggested as an explanation of the differentiation of the subunits.

Box 6.2 **Calculations of gene flow by F-statistics: a numerical example**

In this numerical example, three sites are studied. Each site population has 3 subunits. The basic data are the frequencies of the marker – **p** – in each of 3 sites are as follows:

	site A	site B	site C
	0.20	0.15	0.5
	0.35	0.40	0.23
	0.32	0.53	0.47
Average	0.29	0.36	0.40

Calculated heterozygosity for site A follows:

$$H_1 = 1 - 0.2^2 - 0.8^2 = 0.32$$
$$H_2 = 1 - 0.35^2 - 0.65^2 = 0.455$$
$$H_3 = 1 - 0.32^2 - 0.68^2 = 0.4532$$

Average for site A (Hs A) = 0.4034

Estimating gene flow between subunits at site A:

H_{total} (calculated from the average **p** values at site A): $Ht\ A = 1 - 0.29^2 - 0.71^2 = 0.4118$

$$F_{st} = (H_t - H_s)/H_t = (0.4118 - 0.4034)/0.4118 = 0.02$$
$$Nm = (1 - 0.02)/4 \times 0.02 = 12.25$$

Conclusion for site A: Only 2% of the variance is among subunits at this site. This can be due to extensive gene flow between them – 12 migrants exchanged every generation. (similarly calculate for sites B and C):

Site B: $(H_s B) = 0.4111$; $(H_t B) = 0.460$; $F_{st} = 0.106$; $Nm = 2.11$

Site C: $(H_s C) = 0.4501$ $(H_t C) = 0.480$; $F_{st} = 0.061$; $Nm = 3.84$

Variation among sites:

Average allele frequency: (0.4 + 0.36 + 0.29) / 3 = 0.35

$$H_s = (0.4034 + 0.4111 + 0.4501)/3 = 0.4215$$
$$H_t = 1 - 0.35^2 - 0.65^2 = 0.455$$
$$F_{st} = (0.455 - 0.4215)/0.455 = 0.0736$$
$$Nm = 3.15$$

About 7% of the variance in this example is contributed by differences among the site means. On average, 3 migrants are exchanged among sites every generation.

The advantage of Wright's F statistics is that they can be calculated from experimental data on the frequency of marker alleles (Box 6.2). Two quantities are calculated from the marker frequencies: the total heterozygosity (H_t) and the average heterozygosity (H_s). From these, F_{st} is calculated as

$$F_{st} = (H_t - H_s)/H_t$$

Heterozygosity (H) is calculated for a site with 2 or more alleles as $1 - \Sigma p_i^2$, where the summation is over all alleles at the locus (in many cases, heterozygotes cannot be directly identified, and H is therefore calculated assuming Hardy-Weinberg equilibrium, as in the example below, Box 6.2). For 2 alleles, $H = 1 - p^2 - q^2$.

THE METHOD OF PRIVATE ALLELES

A second method for estimating gene flow among populations was proposed by Slatkin (1981). The method makes use of the frequencies of rare alleles (rare alleles are those which are present at low frequency). A marker allele which appears in only one subunit (population) is referred to as a **'private allele'.** For estimating gene flow by this method, two kinds of data are needed: the number (or proportion) of subunits in which the marker allele is present, and its average frequency in those subunits.

Rare alleles which appear in many subunits are evidence for gene flow – because independent mutations in different populations are unlikely. On the other hand, rare alleles with limited distribution but high conditional frequency indicate isolation and restricted gene flow (examples are given by Slatkin, 1981). Using computer simulation, Slatkin found a linear relationship between the average frequency of private alleles (p_1) and **Nm**, depending on population size. In the case that $N \sim 25$,

$$\ln \mathbf{Nm} = (\ln (p_1) - 2.44)/0.505$$

e.g., if $p_1 = 0.3$ then **Nm** ~ 11.6.

Examples

The efficacy of the two methods for estimating gene flow in natural populations was tested in many studies. For example, Costa and Ross (1994) studied the differentiation of a moth which infests black cherry (*Prunus serotina*) along the east coast of the USA using F-statistics. They found low values of F_{st} and a relatively high gene flow (**Nm** = 4 – 14) although the populations were at least 100 km distant from each other. At the other end of the geographical scale, Schilthuizen and Lombaerts (1994) studied the snail populations (*Albinaria conugata*) on the island of Crete, which were 10 m apart. Although the measured mobility of the snails is limited (maximum 2 - 4 m per year), **Nm** values still indicated gene flow (4.5–17.6 individuals exchanged per generation). The observed relationship between geographical distance and dispersal ability was reviewed by Peterson and Denno (1998).

Values of **Nm** < 1 indicating genetic differentiation due to random drift, were reported in salamander populations along streams in Kentucky, USA (Storfer, 1999). The factor limiting gene flow was predation by fish on the young salamanders as they tried to move between sites!

Martinez et al. (2005) used F-statistics to estimate gene flow among populations of a gall-inducing aphid (*Baizongia pistaciae*). This is a host-specific aphid (in Israel) which galls *Pistacia palaestina*. Two or three trees were sampled in each of four sites, and 10-20 galls were collected per tree. Forty-one genetic markers were identified by molecular methods in aphids of each gall, and **F_{st}** was calculated for each marker as in Box 6.2. Although the aphids can fly and do disperse from the galls during two stages in their life cycle (review in Wool, 2004) the result showed that trees which are located 200 m from each other at a site carry genetically distinct aphid populations (Nm < 1) (Table 6.1).

Table 6.1 Gene flow in a galling aphid, *Baizongia pistaciae*, on *Pistacia palaestina* (from Martinez et al., 2005). A site where the sampled trees grew close together is highlighted

Site	# trees	intertree distance	F_{st} (SE)	Nm
Avivim	3	300 m	0.38 (0.065)	0.50
Shomera	2	200 m	0.48 (0.076)	0.27
Kiryat Tivvon	**2**	**50 m**	**0.07 (0.020)**	**3.26**
Beit Shemesh	2	200 m	0.24 (0.65)	0.81

The usefulness of Wright's F statistics increases with the growing interest in animal and plant conservation, where decisions are often based on estimates of genetic similarity among populations.

Estimates of gene flow using private-alleles were published by Kourti (2004), in an electrophoretic study of the Mediterranean fruit fly, *Ceratitis capitata*. Samples were obtained from different parts of the world. Surprisingly, estimates of **Nm** among sites of a single country (Greece) were very similar to those obtained among continents **(Nm** ranging between 2.8 and 3.2), although the differences in allele frequencies were very large.

[I took the liberty of using the allele frequencies in Kourti's Table 2, to calculate **Nm** among continents by F-statistics (Table 6.2). The results of using this method are quite different from Kourti's. Of 15 values of global samples listed, four gave **Nm** < 1, and six gave **Nm** between 1 and 2 migrants per generation. Only four values of inter-continental differentiation were as high as those estimated by private alleles. This result makes

Table 6.2 Gene flow calculated from F-statistics among global samples of the medfly, (Source: Table 2 in Kourti, 2004).

Marker	F_{st}	Nm	Marker	F_{st}	Nm
1	0.302	0.578	8	0.185	1.101
2	0.148	1.439	9	0.089	2.559
3	0.118	1.869	10	0.050	4.750
4	0.442	0.250	11	0.143	1.498
5	0.063	3.718	12	0.191	1.059
6	0.345	0.475	13	0.467	0.285
7	0.193	1.045	14	0.418	0.349
			15	0.071	3.271

biological sense, as the distance between continents should be expected to limit dispersal (unless *C. capitata* is dispersed by man).]

Literature Cited

Cockerham, C.C. and B.S. Weir. 1993. Estimation of gene flow from F-statistics. Evolution 47: 855-863.

Costa, J.T. and K.S. Ross. 1994. Hierarchical genetic structure and gene flow in macro-geographic populations of the eastern tent populations (*Malacosoma americanum*). Evolution 48: 1158-1167.

Kourti, A. 2004. Estimates of gene flow from rare alleles in natural populations of the medfly, *Ceratitis capitata* (Diptera: Tephritidae). Bulletin of Entomological Research 94: 449-456.

Kaufman, B. and D. Wool. 1992. Gene flow by immigrants into isolated recipient populations: a laboratory model using flour beetles.Genetica 85:163-171.

Lyell, C. 1873. The Geological Evidence of the Antiquity of Man. 4th edition. Murrey, London.

Martinez, J-J.I., M. Wink and D. Wool. 2005. Patch size and patch quality of gall-inducing aphids in a mosaic landscape in Israel. Landscape Ecology 20: 1013-1024.

Merrell, D.J. 1965. Competition involving dominant mutants in experimental populations of *Drosophila melanogaster*. Genetics 52: 165-189.

Narise, T. 1968. Migration and competition in *Drosophila*. I. Competition between wild and vestigial strains of *Drosophila melanogaster* in a cage and migration tube population. Evolution 22: 301-306.

Peterson, M.A. and R.F. Denno. 1998. The influence of dispersal and diet breadth on patterns of genetic isolation by distance in polyphagous insects. American Naturalist 152: 428-446.

Schilthuizen, M. and M. Lombaerts. 1994. Population structure and levels of gene flow in the Mediterranean land snail *Albinaria corrugata* (Pulmonata: Clausilidae). Evolution 48: 577-586.

Slatkin, M. 1981. Estimating levels of gene flow in natural populations. Genetics 99: 323-335.

Sokal, R.R., A. Kence and D.E. McCauley. 1974. The survival of mutants at very low frequencies in *Tribolium* populations. Genetics 77: 805-818.

Storfer, A. 1999. Gene flow and population subdivision in the streamside salamander *Ambystoma barbouri*. Copeia (1944), 174-182.

Thornton, I.W.B. and N.J. Rosengren. 1988. Zoological expedition to the krakatoa islands, 1984 and 1985. I General introduction. Philosophical Transactions of the Royal Society, London B: 322: 273-316.

Via, S., A.C. Bouck and S. Skillman. 2000. Reproductive isolation between divergent races of pea aphids on two hosts. II. Selection against migrants and hybrids in the parental environment. Evolution 54:1626-1637.

Wallace, B. 1969. Topics in Population Genetics. Norton, New York, USA.

Wool, D. 2004. Galling aphids: specialization, biological complexity, and variation. Annual Review of Entomology 49: 175-192.

CHAPTER 7

Deviations from Equilibrium: Non-random Mating

One of the more important requirements for the Hardy-Weinberg equilibrium is that mating in the population is random, in the sense that every genotype has equal probability to mate with any other (**panmixia** was the term coined by August Weismann at the end of the 19th century). If this requirement is not met, then the exact probabilities of mating between all genotypes must be known before the frequencies of offspring in the next generation can be estimated.

In the absence of evidence to the contrary, the assumption that mating in nature is random appears acceptable: there does not seem to be any reason to assume that mating in flocks of house sparrows, house flies, wind-pollinated plants or mice and rats in human habitats is other than random. But a deeper consideration will show that random mating in nature is rare, if it exists at all.

NON-RANDOM MATING IN NATURAL POPULATIONS

Deviations from randomness can take several forms. One such is a preference in mate selection. The lady beetle *Adalia bipunctata* in England has two morphs – red (typical) and black (melanic). Muggleton (1979) collected 100 mating pairs in copula and compared the frequency of the two morphs in these pairs with a random sample of 603 beetles. While 11% in the random sample were melanic, in the copulating pairs their frequency was 22%. The difference was statistically significant.

Cases of non-random mating can be easily brought to mind by considering well-known patterns of mating in animals like deer, antelopes, monkeys and wolves, which live in herds or packs with a strong hierarchical social structures. It is quite clear that those individuals which actually mate are not a random sample of the entire population, although what it is that makes one individual dominant in mating is not known (Darwin's book on sexual selection (Darwin, 1874) brings together interesting facts on this issue).

In human populations, there are no biological restrictions on mating between men and women from different parts of the world. Still, mating in human populations is far from random – due to geographical, political, religious and social factors. Proof of the effectiveness of such barriers to random mating was obtained in many research papers which measured the genetic similarity between the Jewish population in different countries and the non-Jewish population in the same countries (e.g. Kobyliansky et al., 1982). The Jewish population in most cases was genetically distinct – probably because they did not intermarry with their neighbors, due to religious reasons and imposed restrictions.

In many families of plants, self-fertilization in the same flower is common. Many families of insects and other invertebrates reproduce parthenogenetically or otherwise asexually. In these cases, of course, no random mating can occur.

(Some genetic characters can be treated as irrelevant to mate choice, and therefore expected to vary at random – such as blood-group characters or the ability to taste phenyl thiocarbamate (PTC). Although direct selection is probably not involved, these characters may not behave randomly because of selection on other characters of the same individuals.)

INBREEDING AND OUTBREEDING

If mating is not random, prediction of genotype frequencies in the offspring is possible only if the mating system can be specified. Mating systems can roughly be classified into two groups: **inbreeding systems** in which matings between relatives are more frequent than expected at random – like selfing or mating between cousins – and **outbreeding systems**, in which the mating of relatives is discouraged, avoided or prevented – like self-incompatibility in plants. The genetic consequences of these mating systems are very different.

Since there is a variety of mating systems, the discussion is facilitated by looking at the extreme cases in these two classes. The extreme inbreeding is **selfing** – where each genotype mates exclusively with its own genotype (e.g., fertilization within the same flower in plants). Extreme outbreeding is **self-sterility** or **self-incompatibility**, as discovered in the plant family Primulaceae by Charles Darwin (1877). Understanding the consequences of these two extremes will enable at least a qualitative assessment of the expected outcomes in intermediate cases.

EXTREME INBREEDING: SELFING

Self-fertilization is rather rare in animals but quite common in some families of plants. The genetic consequences of this mating system are shown in Box 7.1.

Box 7.1	The genetic consequences of selfing		
		Genotype frequencies	
Generation	AA	Aa	aa
0	-	1	-
1	1/4	1/2	1/4
2	3/8	1/4	3/8
3	7/16	1/8	7/16
4	15/32	1/16	15/32
infinity	1/2	0	1/2

We start with a population at equilibrium – or even with a population composed entirely of heterozygotes, which will reach equilibrium proportions in one generation. In a selfing population, the homozygous genotypes will continue to reproduce AA or aa. The heterozygote Aa, when selfed, will segregate all three genotypes: ¼ AA and ¼ aa will be added to the homozygote classes. The frequency of the heterozygote Aa will be reduced by half every generation, and eventually become zero:

$$H_n = 1/2 H_{n-1} = (1/2)^n H_0$$

SIB-MATING

Mating between brothers and sisters is a less extreme form of inbreeding than selfing. It is practiced frequently in breeding of domesticated animals and in laboratory experiments. Calculation of expected frequencies under sib-mating is more complex than in random mating.

The British geneticist J.B.S. Haldane noticed that in sib-mating, four types of mating combinations are possible ('families') which he labeled W, X, Y and Z (Fig. 7.1).

Parents

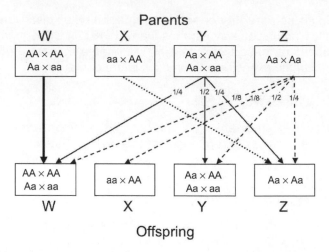

Offspring

Fig. 7.1 Segregation of genotype 'families' in sib-mating populations (Haldane's method, after Li, 1956).

W: homozygous families: AA × AA or aa × aa, which continue to produce their own genotypes.

X: Families of two different homozygous genotypes: AA × aa . All their expected offspring will be Aa (which will join the Z class).

Y: Families where one parent is heterozygous and the other homozygous: Aa × AA or Aa × aa. The offspring – half of them heterozygotes in each cross – will join the appropriate classes.

Z: Families where both parents are heterozygous, Aa × Aa. These families will yield offspring in all 4 groups.

Haldane showed that the frequencies of the four kinds of families can be predicted from four recurrence equations. Knowing their frequencies at any generation n, the frequency at generation n + 1 is given by

$$W_{n+1} = W_n + 1/4\ Y_n + 1/8\ Z_n$$
$$X_{n+1} = 1/8\ Z_n$$
$$Y_{n+1} = 1/2\ Y_n + 1/2\ Z_n$$
$$Z_{n+1} = X_n + 1/4\ Y_n + 1/4\ Z_n$$

Since the output of each mating family is predictable from Mendelian segregation, so too are the total genotype frequencies in the next generation (Box 7.2).

The final genotype proportion under sib-mating will be similar to selfing: the entire population is expected to become homozygous, and the heterozygotes will eventually disappear. However, the approach to this final state will be somewhat slower than in selfing:

$$H_{n+1} = 1/2\ H_n + 1/4\ H_{n-1}$$

Box 7.2	Expected genotype frequencies in the offspring of sib-mating families at a marker locus with 2 alleles		
Generation	AA	Aa	aa
0	-	1	-
1	1/4	1/2	1/4
2	1/4	1/2	1/4
3	5/16	3/8	5/16
4	11/32	5/16	11/32
5	24/64	8/32	24/64
infinity	1/2	0	1/2

Changes in Allele Frequency during Inbreeding

It should be emphasized that inbreeding – even the most extreme inbreeding, selfing – does not by itself change <u>allele frequencies</u>, as long as no other factor, like genetic drift or selection, affects the population, a single generation of random mating will restore genotype frequencies to the original equilibrium frequencies even after many generations of inbreeding (the reader may look at Box 7.1 and confirm that at any generation, allele frequencies remain p = q = 0.5). This is one of the explanations for the prevalence of inbreeding in natural populations.

Wright's General Equilibrium Formula

If the parents are related, the two gametes uniting to form a new zygote (organism) will carry more or less similar genetic information – depending on the degree of relatedness of the parents. If we measure this similarity quantitatively, it will be expressed as a positive correlation between them. The American geneticist Sewall Wright labeled the correlation between uniting gametes the **inbreeding coefficient, F.** He showed that the Hardy-Weinberg equilibrium proportions are a special case of a more general law, such that at equilibrium

$$D = p^2 + pqF$$
$$H = 2pq\,(1 - F)$$
$$R = q^2 + pqF$$

When mating is random, F = 0 and we get the familiar Hardy-Weinberg frequencies. Inbreeding causes an increase in the frequency of homozygotes and a corresponding decrease of heterozygote frequencies, depending on the degree of relatedness of the parents (Fig. 7.2). F can be calculated at any generation from genotype frequencies. As shown before,

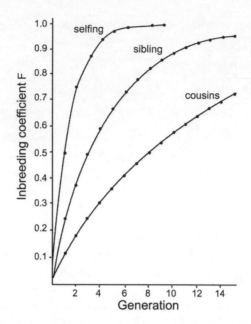

Fig. 7.2 Dependence of the inbreeding coefficient F on the relatedness of the parents (after Strickberger, 1968).

at equilibrium H = 2**pq**, and $H^2 = 4p^2q^2 = 4DR$. With inbreeding, homozygote frequencies will increase and therefore $4DR > H^2$. The inbreeding coefficient is calculated as

$$F = (4DR - H^2)/(4DR - H^2 + 2H)$$

IDENTITY BY DESCENT AND PEDIGREES (THE METHOD OF PATH COEFFICIENTS)

Sewall Wright devised a method by which the degree of inbreeding could be calculated from pedigrees. Inbreeding increases the probability that two gametes, uniting to form a zygote, are identical by descent – i.e., came from the same individual parent one or more generations previously. This identity by descent can be shown in diagrams like Figure 7.3. The immediate result of inbreeding is an increase in the probability of allele identity by descent.

It is possible to generalize the method illustrated in Figure 7.3 to calculate F in more complex pedigrees, as illustrated in Strickberger (1968), and also for other uses (e.g. in Sokal and Rohlf, 1995).

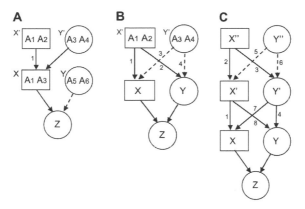

Fig. 7.3 The calculation of the inbreeding coefficient from pedigrees by the method of **path coefficients**. A. The two parents of Z (labeled X, Y) are unrelated. B. The parents are siblings. C. Two generations of sib-mating. Each path coefficient from parent to offspring (arrows) = 0.5. The inbreeding coefficient of the offspring (Z) is <u>half the sum of the products in all paths connecting its parents</u>. For examples, see Box 7.3.

Box 7.3 **Examples of calculations of inbreeding coefficients from pedigrees**

A path from parent to offspring (arrow), or the probability of identity by descent, is 1/2. The inbreeding coefficient of the offspring (Z) is calculated as <u>half the sum of the products of all paths connecting his parents</u>. The paths are indicated by numbered arrows.

A. Only one path will enable Z to get an ancestral allele A_1 from his grandfather. The inbreeding coefficient of Z when the parents are unrelated is $\frac{1}{2} \times \frac{1}{2} = \frac{1}{4}$.

B. The parents X, Y are siblings. Two 2-step paths connect the parents: follow the arrows (1-2), (3-4). The probability that X and Y will both get A_1 is $(\frac{1}{2})^2$, and so is the probability that they both get A_2. The inbreeding coefficient of Z is

$$F_Z = \frac{1}{2} \times (\frac{1}{2}^2 + \frac{1}{2}^2) = 1/4$$

C. This is a more complex network. Six paths connect the parents X and Y: there are four 4-step paths [1-2-3-4; 1-5-6-4; 7-6-5-8; 8-2-3-7)] and two 2-step paths [1-8; 7-4]. Therefore after 2 generations of sib-mating is

$$F_Z = 1/2 \times [4 \times (1/2)^4 + 2 \times (1/2)^2] = 3/8 = 0.375$$

DELETERIOUS EFFECTS OF INBREEDING

"It is hardly an exaggeration to say that nature tells us, in the most emphatic manner, that she abhors perpetual self fertilization" (Darwin, 1877) (1984, p. 293).

Darwin did not know about genes and genetics, but farmers and plant and animal breeders in the 19th century were well aware of the deleterious effects of inbreeding. For example, as early as 1845, a breeder wrote:

"If we shall breed a pair of dogs from the same litter, and unite again the offspring of the pair, we shall produce at once a feeble race of creatures, and the process being repeated for one or two generations more, the family will die out, or be incapable of propagating their race" (G. Low, cited by Wallace, 1889, p.160).

In his book on domestication, Darwin (1898) reviews the information available to him on inbreeding of dogs, horses, cattle, pigs, sheep and pigeons. He himself observed in detail the effects of close inbreeding of fancy pigeons and experimented with inbreeding in plants. It is quite surprising how much information was available on the consequences of inbreeding even before the nature of the phenomenon itself was understood.

Increased identity by descent may have a short-term ill-effect on the offspring. Most organisms carry deleterious alleles in the heterozygous state, of which they are unaware. If parents are related, they may share such alleles – and some of their offspring may be homozygous and express deleterious traits phenotypically. The probability of this occurring will increase the more closely-related the parents are.

That inbreeding has deleterious effects was probably known to people in ancient cultures. Taboos on incest and restrictions on marriage within families are known in many cultures (a recent review of the subject: Aoki 2005) – although such marriages were desirable for retaining the family wealth (the marriage of a daughter to someone outside the family meant giving away some property as dowry). Considerable research was carried out on small and isolated human population, where inbreeding may be prevalent (e.g., Bonne-Tamir et al., 1978a,b). An increase in the frequency of deleterious congenital traits was often found. A famous example is the gene Acheiropedia in some families in Brazil, which causes atrophied arms and feet when homozygous (the heterozygotes are normal. Fraire-Maia and Chakraborty, 1975). The frequency of sib-mating in these families was twice as high as in the population as a whole.

There are indications that infant mortality may be higher in inbreeding families than in the general population. Negative effects of inbreeding were detected in characters such as blood pressure, eyesight and hearing, and IQ (Schull et al., 1970a,b).

Apart from the expression of malformations at the individual level, more subtle negative effects of inbreeding, known as **inbreeding depression,** are often an unwanted – but unavoidable – result of selective breeding programs designed for improvement of some character of interest. Selection necessarily restricts the gene pool by limiting the number of parents to a small number which show the desirable trait. This increases the probability of identity by descent.

There is a growing interest in inbreeding depression in captive populations of wild, endangered species, which are maintained as breeding nuclei in conservation projects, in the hope of eventually releasing them back into in nature. Ralls et al. (1988) and Ralls and Ballou (1988) collected data on the survival of about 40 species of captive marsupials, rodents, carnivores, various antelopes and primates from pedigrees in zoos where the parents of each individual were known. They concluded that an individual born to related parents has a far lower chance of survival than an offspring of unrelated parents of the same species, even in the well-protected environment of the zoo. They suggest that in nature inbreeding may have even more harmful consequences.

Long-term disadvantages of inbreeding

On the evolutionary time scale, inbreeding is disadvantageous because the genetic variation is reduced, making the population less able to adapt and more likely to become extinct should an environmental change occur. It is difficult to find supporting evidence for this statement, since it requires long observation periods and careful monitoring of both the population and the environment. Van Delden and Beardmore (1968) showed that the ill effect caused by inbreeding can be remedied by increasing the genetic variation within the population. An inbred population of *Drosophila* with greatly reduced reproductive output was investigated. Twenty-five males were taken from the population and exposed to x-ray irradiation, a well-known mutagenic treatment. The irradiated males were then returned to the population. The reproductive output was soon restored to normal levels. The mutations caused by irradiation in the sex cells of the irradiated males created new genetic combinations in the gene pool when these males mated with normal females, some of which were expressed in more fertile genotypes.

ISOLATION BY DISTANCE

If a population is subdivided, an overall deficiency in heterozygotes and excess of homozygotes is expected (compared with Hardy-Weinberg equilibrium frequencies) due to variation in allele frequencies among subunits (the Wahlund effect). A similar deficiency of heterozygotes may be caused by the limited dispersal of individuals. Organisms do not purposely travel long distances to find mates, if they can find them nearby. This limited dispersal creates inbreeding. Wright (1943) called this phenomenon, 'isolation by distance'.

Dispersal ability of organisms – especially those that can fly – sometimes seems greater than it really is. In Mark-Release-Recapture surveys, most of the *Drosophila* flies were recaptured less than 100 m from the release point (Wallace, 1969). A recent study of damselflies (Odonata) – rather large flying insects with aquatic larvae, inhabiting marshes with slow-flowing water – showed that most of the adults did not disperse more than 50 m in their lifetime, although some individuals flew as much as 2 km over a rather homogeneous habitat (Watts et al., 2004). The frequency distribution of dispersing damselflies is skewed to the right (Fig. 7.4), and seems characteristic of many other insects.

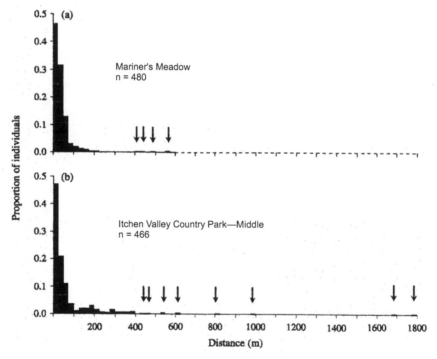

Fig. 7.4 Frequency distribution of dispersal distances of damselflies (after Watts et al., 2004; reprinted with permission of Blackwell Publishing). Arrows indicate rare long-distance flying individuals.

Wind-pollinated plants shed large amounts of pollen, which can potentially be carried over very large distances. Yet careful investigations of domesticated plants, like corn and beet, have shown that pollen dispersal is limited. In corn, less than 18% cross-fertilization was obtained when marked varieties were sown less than 18 m apart. In beet, at a distance of 200 m from a marked variety, the amount of cross-fertilization was only

0.3%. In moth-pollinated flowers, only 2% of the pollen was brought from distances exceeding 200m, although the moths can travel 2-5 km per night to get to the flowers. Similar data prompted Wallace (1969) to describe the structure of insect populations as a fakir's bed of nails: most gene flow occurs within populations, even in the absence of physical barriers.

The magnitude of deviation from expected genotype frequencies due to distance depends on the size of the unit area over which the animals mate practically at random ('**neighborhood size**'). A computer simulation study serves to illustrate the expected effects of isolation by distance. Rohlf and Schnell (1971) created on the computer a population of 10,000 individuals, as an array of 100 × 100 (later 200 × 50 or 1000 × 10). The population was initially seeded by individuals in Hardy-Weinberg proportions with **p** = 0.5, and the genotype of each individual was determined. Genotypes were distributed at random in the array. Each individual in turn was allowed to mate with a randomly-selected neighbor – within a prescribed neighborhood of 9 or 25 individuals (Fig. 7.5).

The expected genotype frequencies among the offspring of each mating pair was then determined from Mendelian segregation, and summed for the entire population at the end of all 10,000 matings. The inbreeding coefficient F was calculated from the formula, $F = (4DR - H^2)/(4DR - H^2 + 2H)$.

A new allele frequency, **p** was calculated at the end of each cycle (generation), and new genotype frequencies determined the genotypes of 10,000 new individuals replacing the old ones at random in the array. The process was allowed to continue for 200 generations. Although mating was

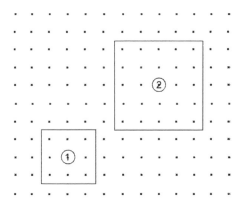

Fig. 7.5 Computer simulation of isolation by distance: neighborhood sizes of 9 and 25 individuals (after Rohlf and Schnell 1971). Mating of the central individuals 1 and 2 was restricted to individuals selected at random from its neighborhood. The frame was centered on each of the 10,000 individuals in turn.

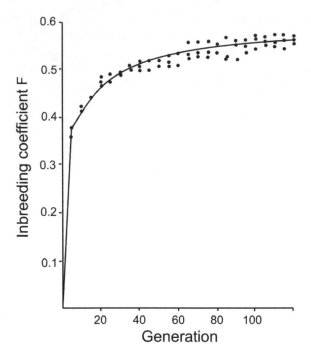

Fig. 7.6 Increase of the inbreeding coefficient F due to isolation by distance. Neighborhood size in this example was 9 (after Rohlf and Schnell, 1971). Dots indicate variation in different runs of the program.

at random between genotypes within the neighborhood, F increased steadily in all simulations (Fig. 7.6). The rate of increase was steeper when neighborhood size was smaller.

VARIATION IN INBREEDING POPULATIONS

Inbreeding – particularly selfing – is expected to reduce genetic variation and cause other ill effects. Nevertheless, natural populations of selfing plants still contain large amounts of genetic variation (e.g. Kahler and Allard, 1981).

Investigation of domesticated plants like oats and barley, in which selfing is almost obligatory, have the advantage that seeds may be stored for long periods and remain viable. Samples of parental plants and their offspring can be planted side by side, or compared electrophoretically on the same gel. An electrophoretic study of esterase variation in seeds of 1500 varieties from the international seed bank showed that 16 of 17 loci were polymorphic, with up to 10 alleles per locus (Kahler and Allard, 1981).

The explanation offered for the maintenance of variation was that homozygotes were selected against in the field, reducing the frequency of deleterious alleles (see next chapter).

Among animals, selfing is common in some species of snails. McCracken and Selander (1980) compared the electrophoretic variation in six selfing species of slugs (*Limax* spp., Gastropoda) with 8 sexually-reproducing species. Almost all the populations of selfers were monomorphic (no variation), as expected, while the sexual species were polymorphic. However, the monomorphic selfers were found to inhabit a greater diversity of habitats than the sexual species! The authors speculate that a single, flexible homozygous genotype with broad physiological tolerance might have been selected. Selfing could be favorable for maintenance of that homozygote genotype.

OUTBREEDING SYSTEMS: SELF-INCOMPATIBILITY

Outbreeding mating systems reduce the probability of union of two gametes identical by descent. Such mating systems are not as common in nature as inbreeding. Darwin thought that the widespread fertilization of plants by insects evolved for promoting cross-fertilization. He considered the structure of orchid flowers, and the elaborate mechanisms for their pollination by moths and bees, as extreme levels of adaptation to cross-fertilization (Darwin, 1877).

The extreme case of outbreeding is found in systems of self-sterility or self-incompatibility. The majority of such cases in multicellular organisms are found in the vegetable kingdom.

Self-sterility was studied and described in great detail by Darwin (1877). In the genus *Primula*, he discovered two forms of flowers in individuals of the same species: a long-styled form with short stamens, and a short-styled form with long stamens. Pollen from each of these forms, if placed on the stigmas, could only fertilize flowers of the other form (therefore in a different individual) ('legitimate' crosses) but not flowers of its own form. This arrangement – called **heterostyly** – was discovered in other species of the same family Primulaceae (e.g., *Lythrum*) and in other families (e.g., in *Oxalis* (Oxalidaceae), where the mating system is even more complex – there are three forms with different relative lengths of styles and anthers, and only six of 18 possible mating combinations are legitimate). This arrangement ensures cross-fertilization, which Darwin was convinced – and demonstrated experimentally – carries great benefits for the plants (Darwin 1891).

Self-incompatibility in diploid organisms requires that at least three alleles – A_1, A_2, A_3 – will be active at a locus determining the mating system.

All individuals must be heterozygous for different alleles at this locus, because A_1A_1 or A_2A_2 cannot exist. The possible (legitimate) mating combinations – and resulting genotypes – are as follows:

Pollen	Stigmas	Resulting genotypes
A_1	A_2 or A_3	A_1A_2, A_1A_3
A_2	A_1 or A_3	A_1A_2, A_2A_3
A_3	A_1 or A_2	A_1A_3, A_2A_3

A self-incompatibility system has other interesting properties. The offspring of each cross will include a genotype which was not one of the parents (e.g., $A_1A_2 \times A_2A_3$ will issue A_1A_2 and $A_1 A_3$: A_2 pollen will not germinate on A_2A_3 stigma). Also important is the fact that any <u>new mutation</u> at the self-incompatibility locus will have a great advantage over all existing alleles, since it will be able to germinate on all stigmas! Thus in a three allele system, A_4 will have initially three times as many chances as the other alleles to fertilize other flowers. This advantage will gradually diminish as more offspring with A_4 accumulate.

As a result of this property, the system will theoretically reach <u>a stable equilibrium</u> where all alleles are at equal frequencies (1/3 for a three-allele system). The initial advantage to any new allele will result in accumulation of many alleles at the self-incompatibility locus. For example, in the mold *Schizophyllum*, there are two loci determining self-incompatibility, each with two subunits. A world-wide survey revealed 32 alleles at locus Aβ and 9 alleles each at Aα, Bα and Bβ (Koltin et al., 1967; Stamberg and Koltin, 1973).

OUTBREEDING DEPRESSION

Outbreeding has been considered favorable, but crossing of widely-distinct varieties may also bring about deleterious effects and loss of fitness (outbreeding depression). This phenomenon is less often mentioned. One case is reported by Goldberg et al. (2005) in fish. Two distinct populations of largemouth bass (*Micropterus salmonides*) were crossed in experimental ponds. The outbred F1 produced 14% fewer progeny than non-outbred fish. More significantly, the F2 progeny were much more susceptible to viral infection: they died 3.5 times faster than the non-outbred fish, and only 35% still survived two weeks after the infection (compared with >80% of the parental stocks).

Literature Cited

Aoki, K. 2005. Avoidance and prohibition of brother-sister sex in humans. Population Ecology 47: 13-19.

Bonne-Tamir, B., S. Ashbel and S. Bar-Shani. 1978a. Ethnic communities in Israel: the genetic blood group markers of the Babylonian jews. American Journal of Physical Anthropology 49: 457-464.

Bonne-Tamir, B., S. Ashbel and S. Bar-Shani. 1978b. Ethnic communities in Israel: The genetic blood group markers of the Moroccan jews. American Journal of Physical Anthropology 49: 465-472.

Darwin, C. 1904 (1877). On the Various Contrivances by which Orchids are Fertilized by insects. 4th edition. Murray, London, UK.

Darwin, C. 1874. The Descent of Man and Selection in Relation to Sex. Hurst and Co., New York, USA.

Darwin, C. 1891. The Effects of Cross- and Self-fertilization in the Vegetable Kingdom. 3rd ed. Murray, London, UK.

Darwin, C. 1898. The Variation of Animals and Plants under Domestication. 2nd ed. Appleton and Co., New York, USA.

Darwin, C. 1986 (1877). The Different Forms of Flowers on Plants of the Same Species. University of Chicago Press, Chicago, USA.

Fraire-Maia, A. and R. Chakraborty. 1975. Genetics of Acheiropedia (the handless and footless families of Brazil) IV. Sex ratio, consanguinity and birth order. Annals of Human Genetics 39: 151-161.

Goldberg, T.L., E.C. Grant, K.R. Inendino, T.W. Kassler, J.E. Claussen and D.P. Phillipp. 2005. Increased infectious disease susceptibility resulting from outbreeding depression. Conservation Biology 19: 455-462.

Kahler, A.L. and A.L. Allard. 1981. Worldwide patterns of genetic variation among four esterase loci in barley (Hordeum vulgare L.). Theoretical and Applied Genetics 59: 101-111.

Kobyliansky, E., S. Micle, M. Goldschmidt-Nathan, B. Erenburg and H. Nathan. 1982. Jewish populations of the world: genetic likeness and differences. Annals of Human Biology 9: 1-34.

Koltin, Y., J. Raper and G. Simchen. 1967. The genetic structure of the incompatibility factors of Schizophyllum commune: the B factor. Proceedings of the National Academy of Sciences, USA 57: 55-62.

Li, C.C. 1956. Population Genetics. Chicago University Press, Chicago, USA.

McCracken, G.F. and R.K. Selander. 1980. Self fertilization and monogenic strains in natural populations of terrestrial slugs. Proceedings of the National Academy of Sciences, USA 77: 684-688.

Muggleton, J. 1979. Non-random mating in wild populations of polymorphic Adalia bipunctata. Heredity 42: 57-65.

Ralls, K. and J.D. Ballou. 1988. Captive breeding programs for populations with a small number of founders. Trends in Ecology and Evolution 1: 19-22.

Ralls, K., J.D. Ballou and A. Templeton. 1988. Estimates of lethal equivalents and the cost of inbreeding in mammals. Conservation Biology 2: 185-193.

Rohlf, F.J. and G.D. Schnell. 1971. An investigation of the isolation by distance model. American Naturalist 105: 295-324.

Schull, W.J., H. Nagano, M. Yamamoto and I. Komatsu. 1970a. The effect of parental consanguinity and inbreeding in Hirado, Japan. I. Still births and prereproductive mortality. American Journal of Human Genetics 22: 239-262.

Schull, W.J., T. Furusho, M. Yamamoto, H. Nagano and I. Komatsu. 1970b. The effect of parental consanguinity and inbreeding in Hirado, Japan. IV. Fertility and reproductive compensation. Human Genetik 9: 294-315.

Sokal, R.R. and F.J. Rohlf. 1995. Biometry. Freeman & Co., New York, USA.

Stamberg, J. and Y. Koltin. 1973. The origin of specific incompatibility alleles: a deletion hypothesis. American Naturalist, 107: 35-45.

Strickberger, M.W. 1968. Genetics. MacMillan, New York, USA.

Van Delden, W. and J.A. Beardmore. 1968. Effects of small increments of genetic variability in inbred populations of *Drosophila melanogaster*. Mutation Research 6: 117-127.

Wallace, B. 1969. Topics in Population Genetics. Norton, New York, USA.

Wallace, A.R. 1889. Darwinism. MacMillan, London, UK.

Watts, P.C., J.R. Rouquette, I.J. Saccheri, S.J. Kemp and D.J. Thomson. 2004. Molecular and ecological evidence for small-scale isolation by distance in an endangered damselfly, *Coenargrion mercuriale*. Molecular Ecology 13: 2931-2945.

Wright, S. 1943. Isolation by distance. Genetics 28: 114-138.

CHAPTER 8

Deviation from Equilibrium: Selection

A population in Hardy-Weinberg equilibrium may be compared to a ball placed on a horizontal surface – like a pool table (Chapter 3). The ball will stay at the same spot as long as the forces which act on it (gravity, friction) balance each other. <u>Selection</u> may be simulated as an external force which pushes the ball away from its former position – when one of the players hits it with his stick: the ball will come to rest again, but not at the same spot. Selection will move the population to a new equilibrium point at a different genetic composition.

In what follows, assume that all the requirements for equilibrium other than selection are met: the population is infinitely large, mating is random, and no mutations or migration takes place. The only force affecting the population is selection. As before, we deal with a single marker locus with two alleles.

BASIC CONCEPTS AND DEFINITIONS

Each genotype – AA, Aa and aa – is assigned a **selection coefficient**, s_i, a quantitative measure of the selective force operating on it. Selection coefficients range from 0 – no selection – to 1, total selection. Usually selection coefficients are positive values.

The value **1–s_i** is referred to as **relative fitness** (or fitness for short) and labeled w_i for genotype i. **W** is the average of **w** in all genotypes (average fitness).

Defined in this way, fitness is a measure of the number of offspring a genotype will leave to the next generation, <u>relative to the best genotype</u> in the population which is assigned a fitness of 1.0. Thus if s_i is positive, genotype **i** will leave fewer $(1-s_i)$ offspring – it is selected against (of course, if s_i is negative, that genotype will have an advantage. This formulation is rarely used). In the absence of selection, the average fitness **W** = 1.

The general concept of fitness is an essential component of the description of selection (for a comprehensive discussion see De Jong, 1994). It is vaguely understood to be related to the success of individuals in the struggle for existence ("fitness is something everybody understands, but no one can define precisely" – Stearns, 1976, cited by De Jong, 1994). w_i is sometimes referred to as the adaptive value of the genotype. This expression is misleading because w_i is not related to adaptation to the environment: it is only a relative measure of reproductive values of the genotypes). It is important to remember that the relative fitness w_i is not a property of individuals, but is assigned to a genotype – as a measure of its relative contribution to the offspring generation.

Selection against a Recessive Genotype

The effect of selection against a deleterious, homozygous recessive genotype **aa** on allele frequencies is shown in Box 8.1. The calculations show that even when the recessive allele is lethal, selection against **aa** will not remove the allele **a** from the population. Of course, if **s** <1 (a deleterious but not a lethal allele), the approach to zero will be even slower. Selection against a recessive allele is inefficient.

The reason becomes clear when we consider the changes in genotype, rather than allele frequencies. Because **q** is smaller than 1, the expected frequency of **aa** (q^2) is much smaller. As selection progresses, fewer and fewer homozygotes will appear in the population, and most of the **a** allele will be carried by Aa heterozygotes which have a normal phenotype and will not be affected by selection (in this model).

Examples of persistence of deleterious and even lethal recessive alleles are well known in human populations. Tay-Sachs disease, and the Acheiropedia gene mentioned above (Fraire-Maia et al., 1975), have disastrous phenotypic effects, but still have not been eliminated by natural selection.

An example from a laboratory experiment with the flour beetle, *Tribolium castaneum*, illustrates how ineffective is selection against recessive homozygote genotypes (Wool and Bergerson, 1979). The homozygous mutant *eu* is characterized by one or more extra pairs of terminal appendages in the larva and the pupa (but not the adult) (Fig. 8.1).

Box 8.1	Selection against the recessive homozygous genotype aa

A. We start with a population in equilibrium (if it is not in equilibrium, a single generation of random mating will get it there).

	AA	Aa	aa
Frequencies before selection	p^2	$2pq$	q^2
Relative fitness w_i	1	1	$1-s$

Average fitness $W = p^2 \times 1 + 2pq \times 1 + q^2 \times (1-s) = p^2 + 2pq + q^2 - sq^2$

$W = 1 - sq^2$ (since $p^2 + 2pq + q^2 = 1$)

B. The effect of selection on allele frequencies in the next generation is calculated from the relation

$$q = R + \tfrac{1}{2}H$$

After selection, $q_1 = (q^2(1-s) + pq)/(1 - sq^2)$

We substitute $p = 1 - q$ to get

$$q_1 = (q(1-qs))/(1 - sq^2)$$

C. Extreme limits of **s**:

When there is no selection ($s = 0$) the last expression becomes $q/1 = q$ (no change in allele frequency). The prediction at the other limit, when $s = 1$ (a lethal allele when homozygous) is not trivial:

$$q_1 = q(1-q)/(1 - q^2) = q(1-q)/(1+q)(1-q) = q/(1+q)$$

To predict the change in **q** for consecutive generations, substitute q_1 for **q** and repeat the calculation. The results are

$$q_2 = q/(1 + 2q); q_3 = q/(1 + 3q); q_4 = q/(1 + 4q) \dots q_n = q/(1 + nq)$$

The series $1, \tfrac{1}{2}, \tfrac{1}{3}, \tfrac{1}{4} \dots 1/n$ will approach zero when n goes to infinity.

The mutant was introduced in generation 0 at a frequency of 0.5 by crossing *eu/eu* **x** +/+. Samples of 100 eggs were taken at the end of each generation to continue the lines, but no artificial selection was applied (mutants cannot be distinguished in the egg stage). The steady decline in mutant frequency (Fig. 8.2) shows that the mutant had much reduced fitness, but it still persisted for 10 generations. Selection was 'natural' in the sense that only interactions between genotypes, and between them and the environment, caused the decline in the frequency of *eu*.

OTHER MODELS OF SELECTION

The formulae of Box 8.1 can be slightly modified to illustrate the effects of other models of selection. For example, selection against a recessive allele

Fig. 8.1 The terminal end of a homozygous *eu/eu* pupa (right) and a normal pupa of *T. castaneum*.

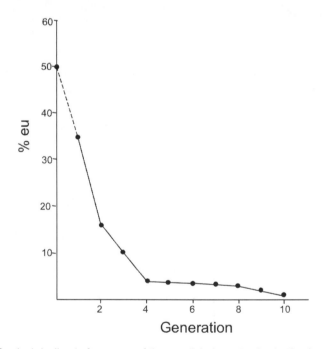

Generation

Fig. 8.2 Gradual decline in frequency of the eu allele by natural selection in a laboratory population of *T. castaneum* (after Wool and Bergerson, 1979).

that also has a negative effect on the heterozygotes can be illustrated by the following model:

	AA	Aa	aa
Relative fitness	1	$1 - s_1$	$1 - s_2$

The prediction of this model (calculations not shown) is that the decline in frequency of the selected allele will be faster than in the preceding case, depending on the magnitude of s_1.

An example of selection against a recessive allele with deleterious effects on the heterozygotes was reported by Dawson (1970) in *Tribolium*. He observed that the decline in frequency of a lethal allele, *Sa*, was faster than predicted theoretically. When negative effects on the heterozygotes were included in the model (s_1 = 0.7 to 0.8, and s_2 = 1), the prediction agreed with the observed results.

Similarly, the model can be set for selection against a dominant allele:

	AA	Aa	aa
Relative fitness:	1 – **s**	1 – **s**	1

Examples of selection against dominant alleles in *Drosophila* were reported by Merrell (1965). Clearly, if **s** = 1 (a dominant lethal allele) the process will be over in one generation.

SELECTION AGAINST HETEROZYGOTES

When the fitness of heterozygotes is lower than that of both homozygotes, a partial reproductive barrier is formed that may lead to the separation of two gene pools. This model of selection is considered a possible mechanism leading to speciation (Box 8.2).

Box 8.2   **Selection against heterozygotes**

The model:

	AA	Aa	aa
Initial frequencies	p^2	2pq	q^2
Relative fitness	1	1-s	1

Average fitness after selection: $W = p^2 + 2pq(1 - s) + q^2 = 1 - 2pqs$

Allele frequency after selection: $q_1 = (pq(1 - s) + q^2)/(1 - 2pqs)$

In the extreme case when **s** = 1 (heterozygotes are lethal – an unlikely event) the equation becomes

$$q_1 = q^2/(1 - 2pq) = q^2/(p^2 + q^2)$$

Allele frequency after selection against heterozygotes depends solely on the frequency of the homozygous genotypes in the parent population in the preceding generation.

[NOTE: in real life, selection against heterozygotes will lead to a change in genotype frequency of one of the homozygotes. Once the initial allele frequency deviates from exactly 0.5, selection against heterozygotes will cause the less-frequent allele to decline continuously until it is entirely eliminated.]

SELECTION IN FAVOR OF HETEROZYGOTES

Heterosis (Hybrid vigor)

When the hybrids between two genotypes have a phenotypic physiological or reproductive advantage over the parents, the phenomenon is described as hybrid vigor. Hybrid vigor is often observed in plant breeding (Darwin, 1891). Farmers and breeders make use of heterosis by crossing different varieties when possible, to get a higher yield.

There are two genetic explanations for hybrid vigor. One is the phenomenon of dominance. Imagine two homozygous parental strains differing at several loci. Each strain carries some favorable alleles (capital letters) and some deleterious alleles. One strain may carry dominant, favorable alleles at loci that are recessive and deleterious in the other. The hybrid will be heterozygous in all loci and express only the 'good' traits:

Parent 1   AA bb CC dd EE ff

Parent 2   aa BB cc DD ee FF

Hybrid   Aa Bb Cc Dd Ee Ff

The second explanation is <u>over-dominance</u>: It is suggested that heterozygotes, which have two alternative genetic factors at their loci, may be more 'flexible' physiologically since they may produce more kinds of enzymes. This may give the individual heterozygote an advantage, e.g., in adaptation to changing environments. However, it is hard to tell which explanation is correct, and each may be right in some cases. Regardless of the mechanism, hybrid vigor can affect the result of selection.

Box 8.3	Selection favoring heterozygotes		
The model :	AA	Aa	aa
Frequencies :	p^2	$2pq$	q^2
Relative fitness :	$1 - s$	1	$1 - t$
Average fitness after selection: $W = p^2(1 - s) + 2pq + q^2(1 - t) = 1 - sp^2 - tq^2$			
Allele frequency after selection: $q_1 = (pq + (1 - t)q^2)/(1 - sp^2 - tq^2)$			
Allele frequencies in the offspring depend on the initial parental frequencies, but also on the selection coefficients **s** and **t**.			

Selective Advantage to Heterozygotes

The consequences of selection favoring heterozygotes are illustrated in Box 8.3. This model has been of particular interest to evolutionists, because when the heterozygotes have a selective advantage, both homozygotes will continue to segregate in the population (even when they are lethal!). Moreover, selection will lead to a stable equilibrium.

At any fixed values of the coefficients **s** and **t** (which are genotype-specific and are supposed to be fixed), plotting **q** over time shows that populations starting from any initial allele frequencies converge on fixed equilibrium frequencies, which are a function only of the selection coefficients! Figure 8.3 illustrates that selection favoring heterozygotes results in a stable equilibrium, which is independent of initial allele frequencies.

The most famous case of selection favoring heterozygotes is the disease sickle-cell anemia in humans. This disease is caused by a point mutation Hb^s in the gene for the β-globin chain in the hemoglobin molecule, replacing a single amino acid (GLU for VAL) (Spiess, 1977). As a result of this replacement, the hemoglobin molecules lose their normal structure and tend to aggregate, losing their function as carriers of oxygen. The red blood cells lose their circular shape, become crescent-shaped ('sickle-shaped'), and form plaques which cause blood clots. The mutation is often lethal in homozygotes. In heterozygotes, some of the blood cells are normal and some 'sickle', causing ill effects.

It was discovered that in areas of the world where malaria is prevalent, the frequency of Hb^s is higher than in non-malarial areas: the 'sickle' hemoglobin seems to be less favorable for the plasmodion, the malaria causer. Mortality of the normal homozygotes from malaria, and of the Hb^s homozygotes from the malfunctioning hemoglobin, gives a selective advantage to the heterozygotes.

A survey of two native populations supports this interpretation. Curacao, an island in the Caribbean, and Surinam – in north-eastern South America, north of Brazil – were both Dutch colonies in the 17th century and received shipments of slaves from Ghana at about the same time. In Surinam, where malaria prevails, Hb^s is as frequent today as in Ghana (~ 0.2). In Curacao, where no malaria is known, the frequency of Hb^s is as low as 0.05. With the selective advantage removed in the absence of malaria, the disadvantage of the homozygotes worked against the mutants.

SELECTION AND AVERAGE FITNESS

If we plot the average fitness, **W**, against allele frequencies when **q** decreases from 1 to 0, in the case of selection against the recessive

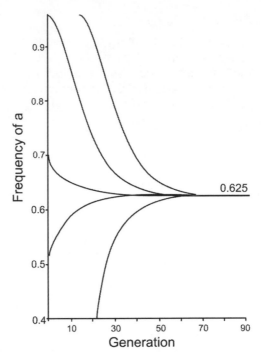

Fig. 8.3 Selection favoring heterozygotes brings about a stable equilibrium. In this example **s** = 0.25, **t** = 0.15. A stable equilibrium is reached at q = 0.25/(0.15 + 0.25) = 0.625, independently of the initial allele frequencies.

genotype, we see that **W** increases as **q** decreases (Fig. 8.4). This makes sense from the Darwinian consideration that the population should be 'improved', on the average, by selection against a deleterious allele.

Similarly, in the case of selection favoring heterozygotes, the average fitness increases with the removal of the deleterious homozygotes. We saw in Box 8.1 that

$$W = 1 - sp^2 - tq^2$$

Substituting **p** = 1 – **q**, we get

$$W = 1 - s\,(1 - q)^2 - tq^2 = (1 - s) + 2sq - (s + t)\,q^2$$

(**s** and **t** are constants). We get **W** as a quadratic function of **q**. Its maximum is where

$$dW/dq = 2s - 2q\,(s + t) = 0$$

at this equilibrium point, **q** = **s**/(**s** + **t**) and **p** = **t**/(**s** + **t**)

Maximum average fitness will be reached at this stable equilibrium point (Fig. 8.5). At allele frequencies lower than the equilibrium point,

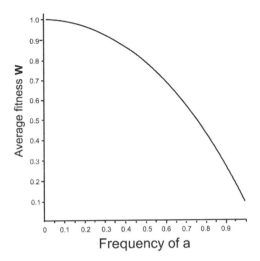

Fig. 8.4 Changes in average fitness **W** when the recessive genotype **aa** is selected against: **W** increases as **q** decreases.

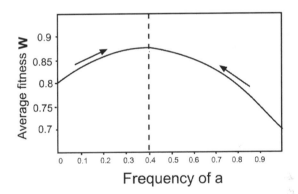

Fig. 8.5 Changes in average fitness **W** in the case of selective advantage to the heterozygotes. The equilibrium point is independent of the initial allele frequencies and is determined only by the selection coefficients.

selection will increase the frequency of **a**. At frequencies above the equilibrium, selection will decrease the frequency of **a** – in both cases selection will invariably increase average fitness.

This general tendency of selection to increase **W** agrees well with the expectation from Darwin's theory of natural selection. Selection is expected to bring about the adaptation of the population to its environment (in the sense of better fitness). True, **W** has nothing to do with the environment: it is only a measure of the average relative reproductive output of the genotypes. But its tendency to increase by selection earned **W** the name 'Darwinian Fitness'.

THE FUNDAMENTAL THEOREM OF NATURAL SELECTION

The behavior of **W** attracted the attention of R.A. Fisher, one of the great theoreticians of population genetics as early as 1930. He showed that the following relation held between genetic variation in fitness in a population – and the rate of change in average fitness (towards equilibrium):

$$dW/dt = \sigma_w^2$$

where σ_w^2 is the genetic variance in fitness in the population. The more variable the population, the faster will it change toward the maximum fitness (equilibrium point):

> "The rate of increase in [mean] fitness of any organism at a given time [ascribable to natural selection acting through changes in gene frequencies] is exactly equal to its genic variance in fitness at that time" (Edwards, 1994).

Fisher's algebraic statement is known as the <u>Fundamental theorem of natural selection</u> (Fisher, 1958). It became very attractive to students of evolution because it appeared to give a theoretical framework to Darwin's belief that selection will improve adaptation.

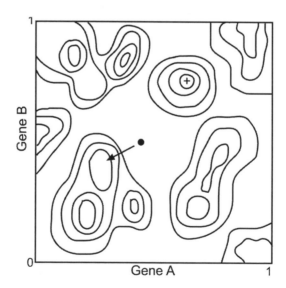

Fig. 8.6 The genetic landscape for two genes, viewed from above. Fitness (**W**) is portrayed as the vertical dimension from the page towards the observer. Contours illustrate levels of fitness. Natural selection can take a population up the nearest peak, but cannot reach a higher peak (e.g., the one marked +) if it involves crossing a valley.

The American population geneticist, Sewall Wright, suggested a graphic illustration of the effects of natural selection on fitness. The **'genetic landscape'** is portrayed as a topographical map. Two genes are illustrated in Figure 8.6. The combinations of their effects result in peaks and troughs of fitness (the fitness vector is vertical from the page towards the observer). From any initial starting point, natural selection will move the population up the <u>nearest</u> peak – not necessarily the highest possible one, because natural selection cannot cause a decrease in fitness (crossing a 'valley'). To reach the highest peak, other forces must intervene – like genetic drift or mutation.

[There is an inherent difficulty in the quantitative approach to selection: it assumes that **s** and **t** are constant properties of the genotypes. But genotypes are not fixed entities: they change every generation by recombination. The order of genes is reshuffled, and the genetic markers do not represent the same genotypes in consecutive generations.]

Literature Cited

Darwin, C. 1891. The Effects of Cross- and Self-fertilization in the Vegetable Kingdom. 3rd ed., Murray, London, UK.

Dawson, P.S. 1970. Linkage and the elimination of deleterious mutant genes from experimental populations. Genetica 41: 147-169.

De Jong, G. 1994. The finess of fitness concepts, and the description of natural selection. Quarterly Review of Biology 69: 3-29.

Edwards, A.W.A. 1994. The fundamental theorem of natural selection. Biological Reviews 69: 443-474.

Fisher, R.A. 1958. The Genetical Theory of Natural Selection. Dover, UK.

Fraire-Maia, A., W-H. Li and T. Maruyama. 1975. Genetics of Acheiropedia (the handless and footless families of Brazil). VII. Population dynamics. American Journal of Human Genetics 27: 665-675.

Merrell, D.J. 1965. Competition involving dominant mutants in experimental populations of *Drosophila melanogaster*. Genetics 52: 165-189.

Spiess, E.B. 1977. Genes in Populations. Wiley, New York, USA.

Wool, D. and O. Bergerson. 1979. Analysis of selection processes using the incompletely-penetrant mutant eu of *Tribolium castaneum*. Canadian Journal of Genetics and Cytology 21: 405-415.

PART II

Selection in Nature

CHAPTER 9

The Theory of Natural Selection: A Historical Outline

PAVING THE WAY

The idea that species developed gradually (evolved) was expressed, published and debated in scientific circles long before Darwin. As stated in his Autobiography (Barlow, 1958), Darwin started collecting data "on the question of species and varieties" in 1837. At the time, the idea of evolution – in the sense of the 'theory of descent' (Haeckel, 1876) – was already 30 years in print. The great French biologist, J.B. Lamarck, in his book, "Zoological Philosophy" (1914; Philosophie Zoologique, 1809) argued that all organisms descended from a common source, and evolved from simple to more complex forms due to the the needs of existence in different environments (see Chapter 1). Lamarck suggested that species change from one form to another – an idea referred to as 'transmutation'. This idea was debated between supporters and dissenters in his native France and in England.

When young Darwin set out for the voyage around the world on board the 'Beagle', in 1832, he carried with him a copy of the first edition of the monumental book, "The Principles of Geology" by Charles Lyell. Lyell, who later became a close friend of Darwin's, introduced two concepts of immense importance into geology. First, that all changes in the surface of the earth – which are recorded in the geological layers of rocks – resulted from the activity of the same forces which are in operation today: volcanic activity, elevation of mountain ranges and erosion. There was no need to invoke extraordinary catastrophes, as advocated by Lamarck's great adversary, Georges Cuvier. Second, that the geological processes required vast periods of time – much more than the few thousand years preached by the Church. These ideas admittedly affected Darwin's thinking.

In 1838 Darwin became familiar with another influential book, 'An Essay on the Principle of Population', by Thomas Malthus (1798). Endeavoring to explain the economic problems of food availability, poverty, misery and crime in English society, Malthus criticized as unrealistic the optimistic ideas of an ideal, all-equal society, which reached England from post-revolution France, and introduced the concept of <u>struggle for existence</u>. Food is produced in direct relation to the cultivated area and increases arithmetically, but populations can grow geometrically. There must necessarily be a conflict between population growth and food availability, a conflict that causes starvation, misery and mortality in humans, and loss of seed and low survival in plants and animals.

While collecting data on species and varieties, Darwin corresponded with breeders, gardeners and farmers. A pigeon fancier himself, he knew a great deal about breeding new (sometimes bizarre-looking) varieties of domesticated animals by artificial selection.

Darwin formed his ideas of natural selection on the model of artificial selection, with Lyell's and Malthus's ideas at the back of his mind.

> "At the commencement of my observations it seemed to me probable that a careful study of domesticated animals and cultivated plants would offer the best chance of making out this obscure problem [of the origin of modifications] … We shall thus see that a large amount of hereditary modification is at least possible … and … how great is the power of man in accumulating by his selection successive slight variations" (Darwin, 1898, p. 3).

A SHORT BIOGRAPHY OF DARWIN

The son of a medical doctor, and the grandson of a celebrated doctor, poet and philosopher (Erasmus Darwin), Charles Darwin was expected to study medicine. He spent two years at the medical school in Edinburgh, but left after observing his first surgical operation (as he admits in his autobiography). He was then sent instead to Cambridge to become a clergyman. There he enjoyed two courses – Geology, taught by Sedgwick, and Botany, taught by Henslow – both courses involved field work. He liked horse riding, shooting and collecting beetles.

The British naval ministry (the Admiralty) at the time was equipping a ship (the 'Beagle') for a mission of mapping the shores of South America.

The young but experienced captain of the 'Beagle', Fitz-Roy, requested a companion who could also be in charge of biological and geological collections during the planned 3-year voyage. Upon the recommendation of his Cambridge teachers, young Darwin (then 23 years old) was given the job.

The 'Beagle' sailed in 1832 and eventually circled the globe in 5 years. Whenever possible, Darwin went ashore and made long excursions into the lands they visited, collecting geological and biological materials. Upon his return to England, and after tending to his valuable collections, he married his cousin Emma Wedgwood, bought a house in the village of Down near London, and lived there for 40 years – rarely leaving the village. He was occupied in collecting data by correspondence, studying plants in his garden and greenhouse, and writing books on a wide range of subjects. He had eight children (one daughter died young). Darwin died in 1882, and was buried in Westminster Abbey.

NATURAL SELECTION: DARWIN'S REASONING

Darwin based his theory on three 'facts of nature'.

(1) Potentially, the population of any species of organism – from flies to elephants – can increase indefinitely: if the offspring of every pair survive and reproduce, populations will increase to astronomical size. Many more individuals of every species are born than can survive. The result is an inevitable **struggle for existence.** (Darwin acknowledges in the Introduction to the Origin of Species, that this doctrine was suggested by Malthus.)

(2) The second 'fact of nature' is that unlimited geometrical growth does not occur. No population grows indefinitely (except perhaps the human population). Apart from periods of temporary outbreaks of locusts and other pests, most population sizes seem to be stable in the long run. This means that the great majority of offspring do not reach reproductive age.

(3) Darwin added the third 'fact of nature' – the fact of <u>variation</u> – from his own observations, supported by the abundant data he collected by correspondence. Within every population of (domesticated) animals and plants there exists a lot of variation in many characters. Even the offspring of a single pair vary in strength and in form, and the breeder can select those which suit his needs or his fancy.

Some are stronger, some weaker and disease-prone more than others, some have longer tail feathers or a better sense of smell, or other traits desirable to the breeder.

Adding this fact to the other two, Darwin made the crucial deduction that the same process must be taking place in nature:

> As many more individuals of each species are born than can possibly survive, and as consequently there is a frequent recurrent struggle for existence, it follows that any being, if it vary however slightly in any manner profitable to itself... will have a better chance of surviving and thus be naturally selected.

Nature can do what the pigeon breeder does: weed out the weaker, the sickly and the less productive and allow the healthier, stronger and more productive to survive and reproduce (from this description stems the expression 'survival of the fittest', coined by the philosopher Herbert Spencer and adopted by Darwin in the later editions of the Origin of Species).

Darwin called this process **natural selection**. He thought that it should make populations better adapted to the environmental conditions prevailing in their habitat. With time, natural selection will accumulate small, heritable changes in different populations. They will diverge slowly from each other and eventually become different species.

Darwin thus adopted Lamarck's idea of community of descent in the animal kingdom, as well as the principle of transmutation – one species can, with time, change slowly and become one or more different species (**'descent with modification'**). Darwin's novel contribution to the science of biology was the mechanism by which these changes could take place – natural selection. Darwin provided the key to the understanding of evolution.

Darwin's wide knowledge of nature, his experience as a world traveler, and his ability as a scientist are reflected in his many books. He described natural selection – not as 'nature red in tooth and claw' but as the end result of the differences in survival and reproductive success of individuals. He recognized that such differences may cancel each other at different developmental or life history stages, e.g., in insects. He recognized an important difference between artificial and natural selection: while the pigeon breeder can only select his stock on the basis of characters he can see and measure, natural selection often works on traits invisible to man. And he knew of the phenomenon of 'correlated responses' – when selection on one trait causes changes, often undesirable (to man), in other traits.

Darwin attributed the evolution of the diverse life forms on earth to the action of natural selection. The offspring of common progenitors, if exposed to different environments, will ultimately evolve into different species, each maximally adapted to the requirements of its environment.

Darwin was impressed by the variation he observed among birds in the Galapagos islands:

> "The most curious fact is the perfect gradation in the size of the beaks in the different species of *Geospiza* ... There are no less than six species, with insensibly gradated beaks. Seeing this gradation and diversity of structure in one small, intimately related group of birds, one might really fancy that from an original paucity of birds in this archipelago, one species has been taken and modified for different ends" (Darwin, 1937; The Voyage of the Beagle, 1845).

This phenomenon is referred to as '**adaptive radiation**' (Darwin did not use this term, which seemed to have been coined by Osborn in 1910; see J. Huxley, 1942, p. 486). The group of birds nicknamed 'Darwin's finches' (Fig. 9.1) have been the subject of intensive research by Lack (1947) and more recently by P. and R. Grant, who have published many articles on recent changes in their populations (see in later chapters).

THE HISTORY OF 'THE ORIGIN OF SPECIES'

Darwin began collecting data 'on species and varieties' with no specific hypothesis in mind (as admitted in his autobiography), but gradually began

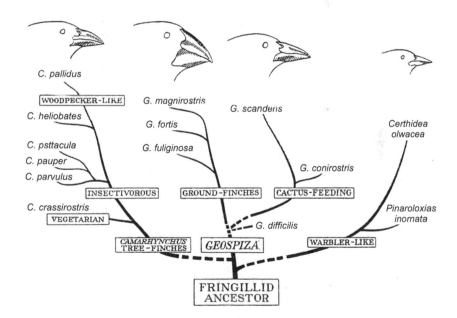

Fig. 9.1 Adaptive radiation of Darwin's finches (from Lack, 1947; reprinted with permission of Cambridge University Press).

to form ideas about descent and selection. He discussed these ideas with his friends – who urged him to publish them. But he was reluctant to do so, for fear of being criticized, and feeling that he had insufficient data to support his hypothesis. By 1844 he had written an abstract – a 'short' draft (240 pages) of the much larger book that he had in mind – and let his friends, Charles Lyell and the botanist Joseph Hooker, read it. Both of them recommended that he should publish the book, to establish his priority over these ideas. Still he hesitated. (Always complaining of ill health, he wrote a will in which he allocated £400 to his wife, asking her to find a good editor and publish 'the book' after his death.) In 1857, he described his ideas on natural selection in detail in a long letter to the American botanist Asa Gray – but asked him not to say anything about these ideas to anyone.

An unexpected event, in June 1858, forced him to to publish the book. He received a short manuscript, accompanied by a letter, from Alfred Russel Wallace – then an almost unknown collector of butterflies and birds in the jungles of the Malay archipelago (now Indonesia). The manuscript – called 'the Ternate paper' after the island where it was written – described the process of natural selection exactly as Darwin had formulated and maintained in his mind for 20 years. In the letter, Wallace asked Darwin to advise him whether his paper was worth publication, and also to show it to Lyell. (In his reply to Wallace, Darwin acknowledged that Wallace was thinking in the same way as himself: "I can plainly see that we have thought much alike and to a certain extent have come to similar conclusions".)

Darwin was shocked to realize that he might lose his claim for priority over his ideas, but showed the paper to his friends Lyell and Hooker as Wallace requested. His friends found a creative solution which established Darwin's priority: Lyell and Hooker presented Wallace's paper to the Linnean Society, together with an abstract of Darwin's 'short' book (which they had read in manuscript 14 years earlier), and a copy of Darwin's letter to Asa Gray. This presentation – in the absence of both Wallace and Darwin – was recorded in the official Transactions of the Society. Darwin set to work on his book, and one year later (1859), 'The Origin of Species' was finally published (the first edition was sold out in one day) – and became a turning point in the history of biology.

ALFRED RUSSEL WALLACE: BIOGRAPHICAL NOTES

Alfred Russel Wallace (1823-1911) had to earn his living from an early age, first as an assistant carpenter, then as a surveyor. He worked in the latter profession for a few years, then became an English teacher, but resigned to take care of his deceased brother's financial matters. In 1844, he traveled with his friend Walter Bates to the Amazon jungles in Brazil, where he collected insects and birds for British collectors. After four years he boarded a ship to return to England, but the ship caught fire and sank with all his collections. He set off again, this time to the Malay archipelago, where he spent eight years.

Wallace, like Darwin, read Lyell's Principles of Geology and Malthus's 'On Population'. He also read Darwin's book on the Voyage of the Beagle. In 1855, he published a short paper, expressing the idea that species are related by descent. In 1858, while lying in his hut feverish with malaria, he pondered on the struggle for existence, and "There suddenly flashed upon my mind the idea of survival of the fittest – that the individuals removed must be on the whole inferior to those that survive". For the two hours of the malaria attack he thought out the details, and the same evening he wrote his manuscript and sent it to Darwin by the earliest ship (which took a few months to reach England).

Wallace described his contribution to the theory of evolution in the following words (Wallace, 1891, p. 21):

> "The immediate result of my paper was that Darwin was induced to prepare for publication his book on the origin of species, in the condensed form in which it appeared, instead of waiting an indefinite number of years to complete a work on a much grander scale. I feel much satisfaction in having thus aided in bringing about the publication of this celebrated book, with ample recognition by Darwin himself of my independent discovery of "natural selection."

Wallace survived Darwin by about 30 years and became a major fighter for the theory of evolution, publishing a number of books and numerous articles in support of 'Darwinism' (the title Wallace chose for his major book).

Wallace disagreed with Darwin on the inclusion of man's moral sense and his mathematical, artistic and musical faculties – among the characters shaped by natural selection. He also disagreed with Darwin on the interpretation of color patterns in insects and birds (the alternative explanations are presented by Darwin in his book on Sexual Selection).

SUPPORT FROM EMBRYOLOGY FOR THE THEORY OF DESCENT

In the 1820s, the German zoologist Karl Ernst von Baer (1792-1876) studied the embryos of different vertebrates. He noted that in early stages of embryogenesis, the embryos of vertebrates of different classes are indistinguishable, while the adults are as different as fish and man. He explained that the traits distinguishing the class are expressed first, then the characters distinguishing family, and finally species characteristics. The more closely-related two organisms are, the more similar will be their embryos at the same developmental stages. For example, the embryos of two species of pigeons will resemble each other more closely than embryos of more dissimilar organisms like pigeon and mammal. (Note that von Baer was considering taxonomic affinity, not community of descent.)

Von Baer disagreed with the opinion of his countryman, Johann Meckel, who claimed that the more 'advanced' vertebrates – before reaching their final form – pass through stages like the organisms lower than themselves in the scale of life (for example, that man passes in his development the stage of a fish, an amphibian, and a reptile before reaching the stage of mammal and finally, human). Von Baer wrote that this comparison is invalid: the different embryonic stages of man should be compared with the embryos of the different vertebrates, not their adult forms.

Darwin realized that von Baer's embryological observations could be used to support his theory of descent. He quotes von Baer in the Origin of Species (p. 364):

> The embryos of mammalia, of birds, lizards and snakes, probably also of chelonia, are in their earlier stages exceedingly like each other... For the feet of lizards and mammals, the wings and feet of birds, no less than the hands and feet of man, all arise from the same fundamental form.

Darwin then adds his own observations on the larvae of crustaceans – especially barnacles – on which he had written a four-volume monograph:

> The larvae of most crustaceans, at corresponding stages of development, closely resemble each other, however different the adults may become.
>
> Even the illustrious Cuvier did not perceive that a barnacle is a crustacean; but a glance at the larva shows this in an unmistakable manner.

The similarity of form between early embryos of vertebrates and the adult forms of other organisms prompted Ernst Haeckel (1876) to coin the expression 'Ontogeny recapitulates phylogeny'. Like Meckel, and unlike von Baer, he interpreted the similarity to mean that new structures are added to the old ones as development proceeds, and the human embryo goes through the stages of fish, amphibian, reptile and mammal, before showing human features: ontogeny is a brief and compact repeat of the phylogenetic history of the species. Due to Haeckel's admiration of and friendship with Darwin, and his prestige as the chief promoter of Darwinism in Germany, Haeckel's (erroneous) interpretation of embryological similarity prevailed for a long time.

THE 'STRUGGLE FOR EXISTENCE' OF THE THEORY OF EVOLUTION

The idea of natural selection and evolution was not acceptable to many in 19th century England. At the front of the opposition were, as could be expected, the clergy, including the leaders of academic institutions like Cambridge and Oxford (which were religious institutions). They were upset by the denial of Creation and the substitution of natural selection for God as the driver of the world and maker of species – and particularly the placement of Man among other animals, and his descent from the apes.

A notable event, perhaps characteristic of the controversy, which left its records in the press of the period and in books about evolution, is described in the 'Life and Letters' of Darwin and of his friend and supporter, Thomas Henry Huxley (see below). It occurred during the annual meeting of the British Zoological Society in the Museum of Zoology in Oxford, in 1860. On the agenda was a discussion of Darwin's book, which had been published the year before, and the main speaker was the Bishop of Oxford, Samuel Wilberforce, who was known as a good speaker and who had promised to crush the theory (Darwin himself did not attend the meeting).

The accounts in the 'Life and Letters' were published many years after the event, and were based on eye-witness memories, since no record was kept of the proceedings. Wilberforce reportedly attacked Darwin's theory as contrary to the Bible. He then cited evidence (on the authority of Richard Owen, a famous Oxford anatomist) that the human brain was fundamentally different from that of the gorilla, and therefore man could not be descended from the apes. Finally he turned to Huxley and asked him, sarcastically, which side of his family was related to the apes – his grandfather's or grandmother's?

There are two or three versions of Huxley's reply (Huxley himself did not remember his exact words when asked). All of them agree, however, that Huxley gave a strong retort that gained him a decisive victory in the debate.

Biographical Notes – Thomas Henry Huxley

Thomas Henry Huxley (1825-1895) had only two years of formal schooling. At age 13 he was sent to live with his sister, who had married a medical practitioner in the outskirts of London, and he served his brother-in-law as an apprentice. Huxley was an avid reader, in particular of scientific books. He taught himself German (later, also French). At age 17 he won second prize in a public competition in Botany, offered by the Charing Cross Medical School – which won him a scholarship there. He excelled in chemistry and physiology.

When he graduated he was still too young to register as a doctor. He joined the British Navy and served five years as an assistant-surgeon on board H.M.S. "Rattlesnake", which was surveying the Great Barrier Reef. Among the detailed research he conducted on board the ship ("with the microscope tied to the table in the map room") he discovered the anatomical structure of the Coelenterates. The two papers he wrote on his research were published before he returned to England. (He also met his future wife in Sydney – but they had to wait eight years before he got himself a suitable job and they could get married.)

Huxley quickly became known as a scientist, specializing in comparative vertebrate anatomy and palaeontology. He became a member of the Royal Society and served as its secretary. He was employed on several government committees and became a key figure in establishing science curricula in schools and universities.

When "The Origin of Species" was published, Huxley promised Darwin to fight for his theory, and did so with great skill and enthusiasm, using his ability as a writer and as a public speaker. He became known as "Darwin's Bulldog".

Steven J. Gould, who in 1977 examined the London newspapers and magazines of the 1860s, suggested that the accounts of the famous debate were biased because the editors of the Life and Letters – the sons of Darwin and of Huxley, respectively – were not impartial. Gould's conclusion was that Wilberforce did suffer a defeat in the debate, but not necessarily because Darwin's theory was proved right; rather, because he did not behave like a gentleman but turned to personal insults.

The Bishop had published a 40-page review of 'The Origin of Species' two weeks before the public debate (Darwin read the review and wrote to a friend that it brought out all the weak points in his book). In the review, he does, of course, declare that man cannot be 'a modified ape' and that the theory is contrary to the Scriptures. But Wilberforce criticizes Darwin not only on religious grounds.

For example, he objects to the use of artificial selection as a model for the natural processes. In particular he attacks the idea that variations such as observed in domesticated animals can be selected to become new species,

> Now all of this is very pleasant writing, especially for pigeon fanciers. But what step do we really gain in it at all towards establishing the alleged fact that variations are but species in the act of formation, or in establishing Mr Darwin's position that a well marked variety may be called an incipient species? We affirm positively that no single fact tending even in that direction is brought forward.

Wilberforce makes a strong case that the theory is based on the notion that a mutation advantageous to the individual can arise and be favored by selection. This assumption is not supported by facts. Darwin's examples are from domesticated breeds; this is philosophically invalid: the observed variants are selected because they are useful for man, not for the good of the animals themselves:

> Every variation introduced by man is for man's advantage …there is not a shadow of ground for saying that man's variations ever improve the typical character of the animal as an animal. They do but by some monstrous development make it more useful to [man] himself … hence it is that nature, according to the universal law with monstrosities, is ever tending to obliterate these deviations and to return to type.

Therefore the theory is no more than a fairy tale,

> In the Arabian nights, we are not offended when Amina sprinkles her husband with water and turns him into a dog, but we cannot open the august doors of the reasonable temple of scientific truth to the genii and magicians of romance (Wilberforce, 1860, p. 250).

Another severe critic was the Scottish engineer, Fleeming Jenkin. Jenkin wrote a review of 'The Origin of Species' in 1867, and raised two interesting objections. One was a rejection of Darwin's use of unlimited time, as an argument which allows most improbable events to occur:

> If a breeder can, in six or sixty years, make a new variety of pigeon, is it likely that in 60,000 years it will become something like a thrush? It is like assuming that since a cannon ball travels a mile in a minute, then in an hour it will cover 60 miles and eventually reach the stars.

Worse, Jenkin wrote: even if a new variant did arise which was twice as fit as the rest of the population, it could not maintain its superiority or be selected as Darwin assumed: it will have to mate with the normal, less favored individuals, and its favorable characters will 'blend' and be reduced by half every generation until it will be lost completely. 'Blending' inheritance – i.e., that an offspring's characters are the average of the parents' – was the common belief in the 19th century (before Mendel's work was rediscovered in 1900) and Darwin could neither answer nor ignore Jenkin's criticism: he changed a paragraph in the 4th and following editions of 'The Origin of Species', and emphasized ordinary individual variation, rather than sudden new 'sports', as the origins of new species.

More serious objections to Darwin's theory came from a biologist, St. George Jackson Mivart, who published a book called 'Genesis of Species' in 1871. Mivart devoted his book to questioning whether Natural Selection could be the only explanation for biological phenomena. In contrast with other critics, Mivart accepted the idea of gradual evolution, but thought that natural selection was insufficient – that an additional force, namely a Divine power, was needed to explain the beginning of new characters. Even those characters that eventually became important for the survival of the organism could not have been useful when their first rudiments appeared, and no selection could favor them.

Mivart's book was published almost simultaneously with Darwin's 'The Descent of Man', and Darwin devoted two full chapters in the later editions of 'The Origin of Species' to reply to Mivart's objections.

Finally, Darwin himself was his greatest critic. He recognized and admitted many difficulties, and tried to answer them as best he could – as any reader of 'The Origin of Species' will find out.

Literature Cited

Barlow, N. 1958. The Autobiography of Charles Darwin. Collins, London, UK.

Darwin, C. 1937 (1845). The Voyage of the Beagle. Collier & Son, New York, USA.

Darwin, C. 1898 (1872). The Origin of Species by Means of Natural Selection. 6th ed. Murray, London, UK.

Darwin, C. 1874. The Descent of Man, and Selection in Relation to Sex. 2nd ed. Hurst & Co., New York, USA.

Darwin, F. 1887. The Life and Letters of Charles Darwin. 3rd ed. Murray, London, UK.

Haeckel, E. 1876. The History of Creation. Appleton & Co., New York, USA.

Huxley, L. 1900. The Life and Letters of Thomas Henry Huxley. MacMillan, London, UK.

Huxley, J. 1942. Evolution: the Modern Synthesis. George Allen & Unwin, London, UK.

Lack, D. 1947. Darwin's Finches. Cambridge University Press, Cambridge, UK.

Lamarck, J.B. 1914 (1809). Zoological Philosophy. MacMillan, London, UK.

Lyell, C. 1853. The Principles of Geology. 9th ed. Murray, London, UK.

Malthus, T. 1798. An Essay on the Principle of Population. J. Johnson, London, UK.

Mivart, St. George J. 1871. The Genesis of Species. MacMillan & Co., London, UK.

Wallace, A.R. 1891. Natural Selection and Tropical Nature: Essays on Descriptive and Theoretical Biology. MacMillan, London, UK.

Wilberforce, S. 1860. On "The Origin of Species by Means of Natural Selection, or the Preservation of Favourable Races in the Struggle for Life", by Charles Darwin M.A., F.R.S. Quarterly Review 1860: 225-264.

CHAPTER 10

Genetic Variation in Natural Populations

HISTORICAL NOTES

Long after the publication of 'The Origin of Species', it was still not clear whether natural populations of organisms possess the kind and amount of variability - which Darwinians assumed to be the basis for the operation of natural selection.

Darwin had in mind two kinds of variation. One was ordinary, inter-individual variation in metric characters, like the lengths of different limbs and other structures of the animal body. Breeders of sheep (Wood and Orel, 2001) and other domestic animals, and fanciers of pigeons, like Darwin himself, were well aware of this variation (Darwin, 1898) and used it in selecting their breeds, but some people thought that this variation was not characteristic of natural populations. A.R. Wallace, in his book 'Darwinism' (1889) presented data of variation in morphological characters of birds and mammals from nature to counter such criticisms (Fig. 10.1).

The second kind of variation known to Darwin – and also to his predecessors and contemporaries – were sudden, rare, major changes in morphology, often referred to as 'sports'. Such changes were observed in domestic animals and, in some rare instances, in humans. Some of these changes were deleterious malformations and the affected individuals were regarded as monsters, such as a lady with 'porcupine' skin, or children born with six fingers on their hands. Some 'sports' were used by animal breeders to start new races. A particularly famous case was a lamb with short legs

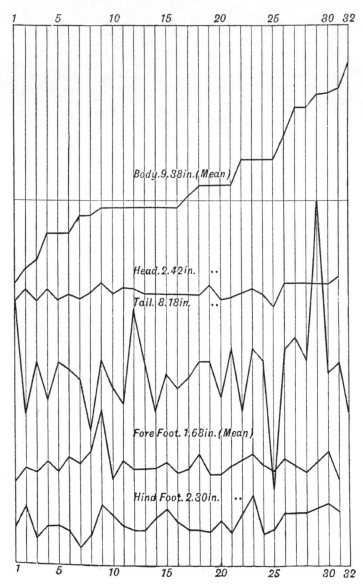

Fig. 10.1 Individual variation in leg and tail lengths of 32 specimens of squirrels (*Sciurus carolinensis*) caught in nature – one of several such drawings in Wallace (1889). Individuals are arrayed by their body size (left to right). The figure illustrates that leg and tail lengths vary independently of animal size.

born in Massachussetts, which gave rise to a breed known as Ancon Sheep.

When Darwin reflected on the appearance of new species, he first thought of such 'sports' as possible origins. But the arguments of Fleeming Jenkin (Chapter 9) convinced him that those 'sports' would not hold their advantage for long in a population. He then changed his emphasis from 'sports' to individual variation as the material used by natural selection to produce new species, and accordingly changed a passage in the later editions of 'The Origin of Species' (Darwin's Life and Letters III: 108-109).

SOURCES OF GENETIC VARIATION

We know today that natural populations of all studied organisms are extremely variable at the molecular (DNA) level. The chance that two individuals are genetically identical is practically zero, except in cases of asexually-reproducing organisms or identical twins.

We also know that only a small fraction of this vast genetic variation is expressed in the phenotypes of the individuals. Until methods were developed in the 1960s which enabled the 'reading' of the gene products, the proteins – and later, the modern DNA sequencing technology – population geneticists could only guess at the amount of variation present, because the phenotypes were all they could measure. The phenotype is affected not only by the genetic content of the organisms but also by non-heritable factors like age, disease and nutrition. **(Nevertheless, it should always be kept in mind that Natural Selection can only detect the phenotypes – and modification of phenotypic proportions may or may not faithfully reflect the changes at the genotypic level.)**

The presence of heritable variation is essential for evolution. Darwin was well aware that a change may only persist if it is heritable, although he did not know how heredity works. No evolution is possible in a genetically homogeneous population. It is therefore important to know how variation is generated and transmitted between generations.

The source of all genetic variation are mutations (Chapter 5) and recombinations, which take place during DNA replication in the meiotic division and the union of gametes (sperm and eggs) in fertilization. Mutations rarely occur, but recombination occurs frequently whenever the sexual cells divide. The phenotypic expression of this variation is restricted, however.

THE 'EPIGENETIC LANDSCAPE' MODEL

To be expressed, the genetic information must first be 'translated' into structural and regulatory proteins. This translation process takes time, and is controlled by feedback mechanisms during the ontogenetic development of the individual [For information on the processes of DNA replication, translation and protein formation, the reader should consult a text like Graur and Li, 2000].

A useful model for understanding the relation between genotype and phenotype was suggested by Waddington (1957). He described the internal environment which controls gene expression during ontogeny as an 'epigenetic landscape', a topographical relief map with ridges and valleys (Fig. 10.2). The length of the map represents developmental time. The fertilized egg (zygote), carrying all the genetic information of the new individual, is portrayed as a ball rolling down one of the valleys. The topography of the landscape directs ('canalizes' in Waddington's term) the genotype to end up in one of the possible phenotypes. The level of 'canalization' differs among characters and organisms.

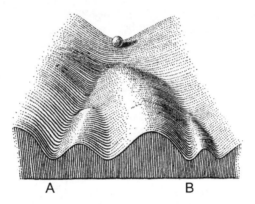

A   B

Fig. 10.2 The epigenetic landscape. The ball represents the zygote, containing the genetic information. The length of the landscape represents ontogenetic developmental time. The landscape directs ('canalizes') the ball to one of the possible phenotypes (A). See text for further explanation (after Waddington, 1957).

Normally, small developmental perturbations (sideways movements of the ball) will not change the resulting phenotype. The ball can be forced into another valley, ending in phenotype B, either by a strong sideways push of the ball (a mutation in real life), or an 'earthquake' lowering the ridge between valleys A and B – a very strong perturbation of the environment. The first is a genetic change, the second is environmental: although it leads to a change in phenotype, it is not heritable.

Canalization may have an indirect evolutionary effect: deleterious mutations which are not expressed in the phenotype are protected from natural selection.

The distance between genotype and phenotype can be visualized, and manipulated experimentally, in cases like the incompletely-penetrant mutation 'extra urogomphi' (*eu*) of the flour beetle, *Tribolium castaneum* (Wool and Mendlinger, 1973). This mutation changes the number of terminal appendages in the larva and the pupa (but not the adult) from the normal two to four (Fig. 8.1) or even six. But the expression of this phenotype is temperature-dependent. At the normal rearing temperature of 30°c, only a few of the homozygous *eu/eu* individuals have the mutant phenotype. Lowering the temperature to 25°c (and consequently increasing developmental time) allows up to 95% of the homozygous mutants to express the mutant phenotype (Wool and Mendlinger, 1973).

Another example of variable expression of a phenotype is the mutant microcephalic (*mc*) in *T. castaneum*. This mutation modifies head size and affects the size and position of the compounds eyes in homozygous *mc/mc* individuals (Fig.10.3). The phenotypic expression seems to be affected by the internal as well as external environment (Wool, 1985).

Fig. 10.3 Variable expression of the microcephalic (*mc*) mutation in homozygous *mc/mc* flour beetles, *Tribolium castaneum*. Top left: a normal individual. Top right: a fully-penetrant mutant (small head, no eyes). Bottom: various grades of expression (from Wool, 1985).

PHENOTYPIC PLASTICITY

Plasticity is the ability to express alternative phenotypes in different environments. Plasticity is not a general property of a genotype: it is specific to a particular trait or group of traits. It is measured, or estimated, when a trait is plotted across environments, in the form called a **'reaction norm'**, one for each genotype. Assigning a reaction norm to a genotype requires that the same genotype be exposed to all tested environments. "A reaction norm as coded for by a genotype is the systematic change in mean expression of a phenotypic character in response to a systematic change in an environmental variable".

Ecologically, a genotype which can survive an unsuitable environment by expressing an alternative phenotype, may have a selective advantage. In the last two decades of the 20th century, the study of variable phenotypic expression of the same genotypes when exposed to different environments attracted much interest, both theoretical and experimental.

From an evolutionary point of view, genetic heterogeneity and phenotypic plasticity are alternative strategies of a species to cope with variation in the environment. Phenotypic plasticity can be an adaptive strategy, pre-empting the need for genetic variation, – or it may be additional to genetic variation (De Jong, 1990).

"Phenotypic plasticity has a retarding effect on the action of environmental selection… it retards selective elimination. It permits a population to survive in an environment that has changed in an unfavorable way…[it] gives a population more chance of acquiring new genetic variation by mutation" (Grant, 1977, p.145).

A great theoretical debate in the literature concerned the question of whether phenotypic plasticity itself is a genetic trait, which can be under selective control (e.g., De Jong, 1990; Scheiner, 1993). De Jong (1995) and Gottard and Nylin (1995), among others, think that it is. To the extent that reaction norms are heritable, the variance among genotypes in reaction norms – measured by some function, like the slope of the reaction norm across environments – may be the material for selection, in the evolution of plasticity (De Jong 1995). In a two-way analysis of variance, with genotypes and environments as main effects, the variation due to genotype × environment interaction is a quantitative measure of plasticity (De Jong, 1995).

Considerable work on phenotypic plasticity was carried out using the African butterfly, *Bicyclus anayana*. Populations of the butterfly were reared in captivity in the Netherlands (Roskam and Brakefield, 1996). This insect has two seasonal morphs in nature, differing in wing pattern and coloration. The environmental factor controlling seasonal morphs was shown experimentally to be the ambient temperature (Windig, 1994). Subpopulations initiated from the same egg batches (families of similar genotypes) were reared under four controlled temperatures. Graphs describing the phenotype of each family in the four ambient temperatures –

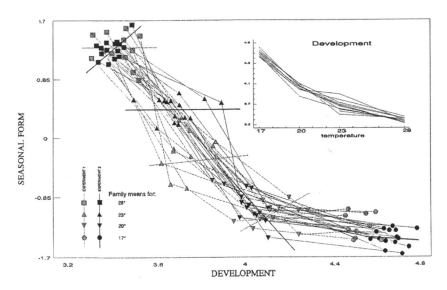

Reaction norms for SEASONAL FORM–DEVELOPMENT time,
for females. Symbols indicate means of families.

Fig. 10.4 Mean reaction norms of families of the butterfly *Bicyclus anayana* reared under four ambient temperatures (inset). Data for each family are connected by a line. Ordinate: an index reflecting developmental time (from Windig, 1994; reprinted with permission of MacMillan Publishers).

their 'reaction norms' – describe the dependence of the phenotype on this environmental factor (Fig. 10.4).

The butterfly samples, although derived from a single egg batch, may still have contained more than one genotype. In the parthenogenetic cotton aphid, *Aphis gossypii*, all replicate families were derived from the offspring of a single female – and have exactly the same genotype. Wool and Hales (1997) obtained aphid samples from localities thousands of kilometers apart in Australia, and reared subpopulations from each in the laboratory on four different host plants: broad bean, cotton, cucumber and melon. When the density in these colonies increased and the host plants deteriorated, alate (winged) aphids appeared in the cultures. These were collected and 25 morphological characters were measured on each individual. The means of the measurements, plotted as reaction norms, describe the performance of the aphid genotypes on the four plants.

The results indicated that the morphological differences among populations, collected from widely distant localities, were much smaller than the difference between aphids (from any single family) reared on the four host plants in the laboratory. The laboratory hostplant had a decisive effect on aphid morphology. Families reared on cotton and broadbean produced smaller alates than on cucumber or melon (Fig. 10.5). In particular, aphids reared on broad bean matured to be very small, and most of them remained apterous ('yellow dwarves'), apparently as a reaction to an unsuitable host plant (Wool and Hales, 1997).

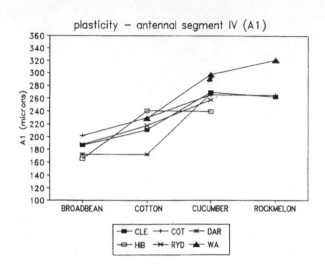

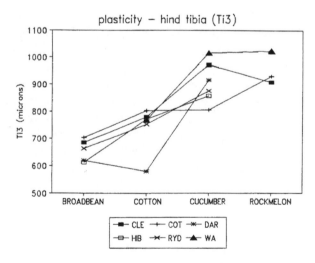

Fig. 10.5 Reaction norms of *Aphis gossypii* when reared on four host plants (environments). Each line represents the mean of a single family (genotype). Two characters are illustrated: antennal segment IV (A1) and hind-leg tibia (Ti3). The four host plants are arranged from left to right by their effect on aphid size, measured as wing length (WTL) (from Wool and Hales, 1997).

WHICH CHARACTERS SHOULD BE USED TO MEASURE GENETIC VARIATION?

In Darwin's time, the known variation in natural animal and plant populations was insufficient to justify Darwin's claim that this could effectively be the source of evolutionary change. Differences in color and size of limbs in animals, flower color and shapes of fruit and seeds in plants, were the basis of the early classification of species by Linnaeus, and the line between variation in such characters among and within species ('varieties') was not clear. The differences observed among individuals within species were quantitatively very small, and it was difficult to imagine that they could be the source for the origin of new species. The acceptance of Darwin's theory depended on the demonstration that heritable variation was abundant.

Variation in morphological characters was known in domestic animals. Such characters were observed and measured, and could be conveniently followed in parents and offspring. But it had to be shown that this variation occurred in nature, and affected animal fitness: in order to claim that natural selection is the driving force in evolution, only such variation that affects the probability of survival and reproduction of their carriers is relevant (Ayala, 1978). Relevant data only became available 50 years or more after Darwin.

In the 20th century, after the discovery of Mendel's paper in 1900 and the subsequent establishment of genetics as a science, variation in many more characters was studied in natural populations. The breakthrough was associated in particular with the introduction of *Drosophila* fruit flies as research animals.

Morphological variation

Hundreds of mutations changing morphological characters like eye-color, wing-shape, and similar traits, were discovered in the 1920s by T.H. Morgan and his group in laboratory populations of *Drosophila.* T. Dobzhansky and his associates conducted large-scale trapping surveys in tropical America between 1930 and 1950 and screened hundreds of thousands of flies from natural populations to estimate the amount of variation in nature. About 2-7% carried morphological aberrations, not all of them heritable. This amount of variation was too small to accept the central role of natural selection in evolution. Moreover, many morphological characters were drastic changes, with deleterious effects on their carriers – such as the loss of wings. Although these variants were heritable and did affect fitness, they could not be considered characteristic of the variation which natural selection could use to create new species.

Chromosomal variation

A number of phenomena at the cellular level were used to monitor the amount of genetic variation in *Drosophila* populations. Some inversions in *Drosophila*, which 'lock up' whole segments of chromosomes and bend the chromosome in a characteristic, recognizable shape, were used as genetic markers. The inversions can be detected under the microscope in the giant chromosomes of the larval salivary glands. T. Dobzhansky and his associates discovered at least 25 such inversions in natural populations of *Drosophila melanogaster* and 12 in *D. persimilis* (Spiess, 1977). The genes associated with one such inversion in *D. pachea*, which breeds in rotting cactus tissues in the deserts of Central America, increase in frequency from southern Arizona to Mexico, suggesting an adaptation to desert conditions (Ward et al., 1974).

Fitness modifiers

The research efforts of Dobzhansky and his group in natural *Drosophila* populations, combined with laboratory breeding and crossing with laboratory strains carrying genetic markers, revealed that most chromosomes sampled from nature carried deleterious alleles, ranging from slight effect to lethality in the homozygote condition. Such genes were found on all chromosomes except the small chromosome Y.

The large proportion of sampled chromosomes which carried fitness modifiers surprised some scientists (Lewontin, 1974). That deleterious alleles are carried in human populations is evidenced by the long list of congenital diseases in man.

POLYGENIC (QUANTITATIVE) CHARACTERS

Many characters have continuous, rather than discrete, phenotype distributions. Some of these characters are very important in animal and plant breeding. Such characters are the lengths of limbs, weight, and yield of different crops – be it milk yield in cattle, egg output in chickens, or wheat yield in fields. These characters fall into a separate category of **quantitative characters.** Their inheritance is treated as a special science, quantitative genetics (Hartl and Clark, 1997).

Historical notes

Quantitative characters must have been on Darwin's mind when he considered variation and its response to selection – simply because this variation was observable in domestic animals and it did respond to artificial selection, as any breeder would testify. Wallace (1889) documented variation in limb length in natural populations of birds and mammals.

Darwin's nephew, Francis Galton, collected data on such characters in humans from pedigrees, which he obtained by sending questionnaires to selected British families – the first attempt at statistical genetics – and attempted to study their inheritance (Galton 1889).

Quantitative genetics separated from evolutionary genetics in the 1900s, with the rediscovery of Mendel's paper. The geneticists ('Mendelians') believed that new species originated in jumps, by mutation. Characters were considered discrete like the mutations they observed in *Drosophila* or the color of flowers in peas. The adherents to the old Darwinian concept of natural selection, on the other hand, argued that species originated gradually. The continuous, quantitative characters were more in line with the Darwinian concept. The rift between the mutationists (geneticists) and the 'biometricians' (Darwinians) held for many years. The theory developed for quantitative characters became the basis of animal and plant breeding, a separate science from population genetics.

The rift between the two genetic theories was bridged by theoreticians like Falconer (1960), who suggested that quantitative phenotypes could be analyzed and interpreted on the same models as classical genetics – only assuming that many genes are involved, each contributing a small amount to the character. Moreover, the phenotypic expression of genes determining quantitative characters is affected by environmental factors: for example, yield depends not only on the genetic characteristics of the plants but also on irrigation, fertilizers etc. Identical twins will differ in height and weight if given different amounts of food.

While 'classical' genetics theory works with single genes at a time, and assumes that the phenotype is an exact reflection of the genotype, quantitative genetics makes no such assumption: it works with the observed phenotypes (and sometimes attempts to estimate how many genes are involved). It "describes phenotypic evolution in terms of parameters that can be measured in natural populations (trait means, genetic variances and covariances, selection gradients) in contrast with the largely unmeasurable gene frequencies and selection coefficients of classical population genetics" (Via and Hawthorne, 2005). The inheritance of such traits can be worked out by applying appropriate statistical tools like the analysis of variance. Francis Galton (1889) made the first attempts in this direction when he applied the normal distribution curve to human measurements data.

INHERITANCE OF QUANTITATIVE CHARACTERS

A characteristic of many quantitative characters – as shown already by Galton – is that if a large number of individuals are measured, and the measurements are arranged from smallest to largest, their frequency distribution approximates the well-known bell-shaped normal distribution (Fig. 10.6).

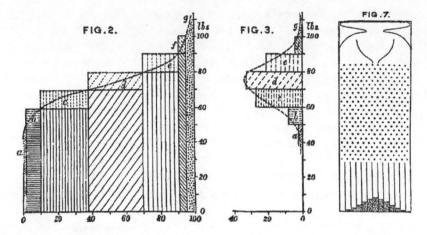

Fig. 10.6 Fit of a normal distribution to large numbers of measurements of a quantitative character. Galton's Fig. 2: cumulative distribution of actual measurements. Fig.3: frequencies in percentages. Fig. 7: an experimental device for illustrating the formation of the curve (from Galton, 1889).

Galton (1889) devised a simple instrument to demonstrate that the combined effect of many random factors results in a normal curve. A large number of small metal marbles are rolled down a slanted board in which many pins are driven at regular intervals. The marbles are collected at the bottom in equally-spaced sectors. Most marbles reach the central sectors, and their density tapers symmetrically towards both ends (a similar device is used today for demonstration in some science museums).

The normal distribution has interesting mathematical properties, and has become the theoretical foundation of most commonly-used statistical tests, such as the analysis of variance, correlation and regression.

In quantitative genetics, the **phenotypic value** of each individual is measured (in units of millimeters, grams, milliliters or any other according to the character under study). In the absence of environmental effects, the expected phenotypic value of a hybrid between two parents is their average (called the **mid-parent value**) but in the case of dominance it may be more similar to one of the parents than to the other.

The quantitative genetic analysis centers on the variance among individuals in the character of interest. (The reader is reminded that the variance is calculated as the average squared difference between individuals and the sample mean in the study.) Falconer (1960) showed that the phenotypic variance can be partitioned into genetic and environmental components,

$$V_p = V_g + V_e$$

In the case of dominance, the genetic component can be further partitioned into an additive and a dominance component:

$$V_g = V_a + V_d$$

(if there is no dominance, $V_d = 0$). The additive variance, V_a, is the important component for inheritance and selection.

How to estimate V_g and V_e?

One method is to compare the marker character - whether measured in nature or in a laboratory experiment – with its value in a control population maintained in a homogeneous environment where V_e is reduced. The difference in magnitude between the two treatments will estimate V_g (usually expressed as a percentage of the total phenotypic variance, V_p). For example, plants collected as seeds from natural populations, if grown in a common greenhouse – thus removing much of the environmental variance - will express mostly the genetic component in the phenotypes.

A field experiment in Israel illustrates that some characters are affected more than others by environmental factors. Samples of seeds of the wild emmer wheat (*Triticum dicoccoides*) were collected at five locations throughout the country and planted at two sites: near Haifa (the Mediterranean climate of Mt. Carmel) and near Sede Boqer, a dry locality in the Negev desert (Nevo et al., 1984). Several quantitative characters were measured. Two are illustrated in Fig. 10.7. The spikelet weight varied among seed populations at both sites, indicating that differences among population means of this variable had a genetic component. But the total plant biomass of all populations was greatly reduced at the desert site: clearly plant biomass had a much greater environmental variance component than spikelet weight (responding to the difference in precipitation between the experimental sites).

Another method of illustrating the genetic and environmental components of the variance is via selection. From a population with mean character value μ_0, select a group of individuals as parents of the new generation from the tail of the distribution, such as $\mu_p > \mu_0$ (as in Fig.10.8).

There are three possible qualitative results. If the mean of the offspring population is equal to the mean of the selected parents, as in B, then the character is fully heritable ($V_e = 0$). If the offspring mean is equal to that of the source population before selection, as in A, then the character has no genetic component and all the observed variation is environmental ($V_g = 0$).

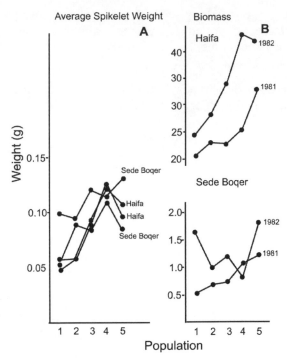

Fig. 10.7 Genetic and environmental effects on characters of wild Emmer wheat. Seeds collected from five populations were sown in a humid (Haifa) and a desert site (Sede Boqer) and quantitative characters of the plants were measured for two years. Spikelet weight (A) shows genetic differences among source populations but little environmental effect of the field sites. Plant biomass (B) reflects the difference in precipitation between the two sites (after Nevo et al., 1984. Note the difference in scale of the upper and lower parts of B).

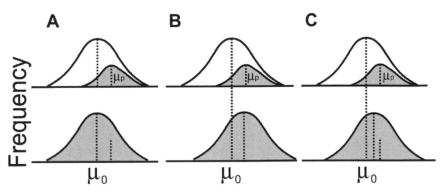

Fig. 10.8 Three possible results of selection on a quantitative character. The distributions of the character in parents (top) and offspring (bottom) populations are compared. A. The variation is entirely environmental ($V_g = 0$). B. The variation is entirely genetic ($V_e = 0$). C. The distribution of the character in the offspring reflects both genetic and environmental effects.

The results of most experiments will be intermediate, as in C – some of the variation is genetic, some environmental. (The normal distributions represent the phenotypic variation around the selected means, which can result from genetic as well as non-genetic causes.)

HERITABILITY

The proportion of the genetic variance, of the total phenotypic variance, is called 'heritability' and denoted h^2:

$$h^2 = V_g/V_p$$

Heritability values range from zero to one. Negative values, and values exceeding 1.0, may be obtained experimentally, but are ignored since a variance (a squared quantity) can only be positive, and V_g cannot by definition exceed V_p.

In selection experiments, heritability can be estimated from a comparison of parent and offspring distributions of a character (Hartl and Clark, 1997). Two variables are calculated: the **selection differential** $S = \mu_p - \mu_0$; and the **selection response** $R = \mu_f - \mu_0$, where μ_p, μ_0, and μ_f are in turn the means of the selected parents, the original population before selection, and the offspring population. When the variables S and R are expressed in standard deviation units, heritability is estimated as

$$h^2 = R/S$$

The concept of heritability is important in practice, because if the heritability of a trait is known, the response of the population to selection can be predicted:

$$R = h^2 S$$

Example

In a quantitative laboratory study of enzyme activity in the flour beetle, Tri*bolium confusum* (Wool and Agami, 2000), each of several males were crossed to three randomly-selected females. The progenies of the individual females were separated and reared to adulthood, and esterase (EST) activity in each parent, and in a sample of individuals from each progeny, was measured colorimetrically. The data were analyzed by a hierarchical ('nested') analysis of variance (Sokal and Rohlf, 1995). The genetic differences among sires (males) contributed much more to the offspring EST activity than the differences among females (dams): enzyme activity in offspring of females mated to the same male were rather similar. Therefore, selection for increased and decreased EST activity was based on the activity in males.

In a long-term experiment, 40 pairs of flour beetles were allowed to oviposit in separate vials to produce 40 offspring families. The activity of EST was then measured colorimetrically in each parent (and later, when the offspring became adults, in samples from each family). The pairs of parents were arrayed in order of magnitude of the male's EST activity. The offspring of the five highest males were selected to continue the H selection line. (Similarly, the five progenies of the lowest EST activity males were selected for the L line.) The procedure was repeated for eight generations.

Mean EST activity increased rapidly in the H line. This was due to the increased frequency of individuals carrying a high-activity isozyme, one of five visible on the gels (Fig. 10.9).

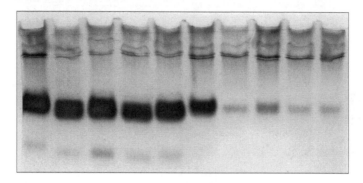

Fig. 10.9 Esterase isozyme patterns in the long-term selection experiment of Wool and Agami (2000). From left: five individuals from the H line, all of them carrying the high-activity isozyme, and five from the L line (one of the five has the high-activity isozyme).

Heritability was estimated every generation from the slope, **b**, of the regression of offspring values on parental EST activity (In regression on values of the male or the female parent, h^2 = **2b;** in regression on the midparent value h^2 = **b.** Falconer, 1960). Experimental values of h^2 ranged from 0.39 to 0.49. Heritability values for many characters in flour beetles (*Tribolium*) range from a few percent to 0.80 (Dawson and Riddle, 1983).

A useful summary statistic for heritability in several generations of selection is '**realized heritability**', calculated from the cumulative response and cumulative selection differential:

$$\text{Realized } h^2 = \Sigma R / \Sigma S$$

The summation is over all generations. In practice, realized heritability is calculated at the <u>last</u> generation of the experiment, as

$$\text{Realized } h^2 = (\mu_f - \mu_0)/(\mu_p - \mu_0)$$

Looked at from a different angle, heritability provides a measure of the magnitude of genetic influences on the character of interest – knowledge of which may indicate the probability of success in selection for that character.

The response of many quantitative characters to selection – as evidenced by the changes in many domesticated plants and animals since man began breeding them – shows that there is a lot of genetic variation in such characters – although we do not know how many genes are involved or how they interact.

THRESHOLD CHARACTERS

Some quantitative characters are expressed as 'all-or-none' binary phenotypes, with no intermediate stages. Among these are characters like resistance to disease or to a pesticide (the individual either survives or dies).

Calculation of the heritability of such characters is complicated because their phenotype is discrete and binary. Characters which have this property are analyzed assuming that the phenotype has an underlying, continuous normal distribution of a hypothetical variable – which may be referred to as 'tolerance' or **liability**. In the case of insecticide resistance, 'tolerance' may depend on the quantity of poison that can be absorbed by the individual (or sequestered by the detoxifying enzyme) before the insect dies (threshold) (Box 10.1). If the level of the treatment exceeds the tolerance level, the phenotype is expressed and the individual dies. Characters that behave like that are called threshold characters (Fig. 10.10).

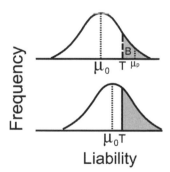

Fig. 10.10 Selection on a threshold character. Top: Parental generation. Bottom: Offspring generation. T is the threshold point. Shading indicates the proportion surviving the treatment. For details refer to Box 10.1. After Hartl and Clark (1997).

What does 'Tolerance' Represent?

The idea that tolerance depends on the accumulation of some physical variable may not be hypothetical. In a series of papers, Moczek and Emlen

---

Box 10.1 **Calculation of the heritability of resistance as a threshold, quantitative character**

If the assumption of a continuous, normal distribution of "tolerance" is accepted as the model, heritability can be calculated from data on observed percentage mortality in two consecutive generations. (Fig. 10.10). T is the threshold point expressed in standard deviation units from the mean. In the normal distribution, there is a theoretical relationship between the height of the curve Z at a given point x, and the mean of the remaining distribution (area) to the right of x. In the selection for a quantitative trait this area represents the selected parents of the next generation. Then their mean is

$$\mu_p = Z / B$$

where B is the proportion surviving the pesticide treatment (area under the tail of the curve). Z is obtained from tables of the normal distribution at the threshold point T (in standard deviation units). From the formula for realized heritability,

$$h^2 = R / S$$

where $$R = \mu_f - \mu_0$$
$$S = \mu_p - \mu_0$$

where μ_0 is the population mean before selection and μ_f is the mean of the offspring generation after selection.

---

studied the expression of frontal horns in the scarabaeid beetle, *Onthophagus taurus*. Males of this small beetle either possess long horns, or do not. Moczek and Emlen (1999) showed that the expression of the horned phenotype is related to the quantity and quality of the food that the female beetle stored for the solitary larva in the dung ball in which she lays her egg. The amount of food determines beetle size. A male must reach a critical (threshold) mass for the horns to be expressed in the phenotype (small males and females, have no horns). (Fig. 10.11).

Literature Cited

Ayala, F.J. 1978. The mechanism of evolution. Scientific American 239: 56-69.

Darwin, C. 1898. The Variation of Animals and Plants under Domestication. 2nd ed. Appleton, New York, USA.

Dawson, P.S. and R.A. Riddle. 1983. Genetic variation , environment heterogeneity, and evolutionary stability. pp. 147-170 in : C.E. King and P.S. Dawson (eds.), Population Biology: Retrospect and Prospect. Columbia University Press, New York, USA.

De Jong, G. 1990. Quantitative genetics of reaction norms. Journal of Evolutionary Biology 3: 447-468.

De Jong, G. 1995. Phenotypic plasticity as a product of selection in a variable environment. American Naturalist 145: 493-512.

Falconer, C.D. 1960. Introduction to Quantitative Genetics. Ronald Press, New York, USA.

Galton, F. 1889. Natural Inheritance. MacMillan, London, UK.

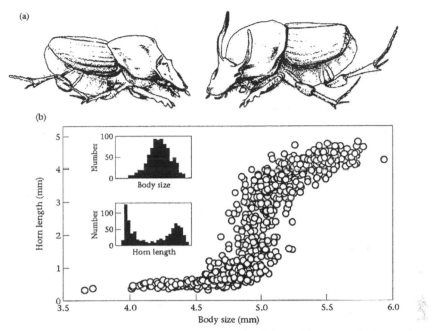

Fig. 10.11 Expression of a threshold character (male horns) in the beetle *Onthophagus taurus* as a function of beetle size, which is determined by the amount of larval food (from Moczek and Emlen (2000); reprinted with permission of the publisher).

Gottard, K. and S. Nylin. 1995. Adaptive plasticity and plasticity as an adaptation: a selective review of plasticity in animal morphology and life history. Oikos 74: 3-17.

Grant, V. 1977. Organismic Evolution. Freeman, San Francisco, USA.

Graur, D. and W-H. Li. 2000. Fundamentals of Molecular Evolution. Sinauer, Sunderland, USA.

Hartl, D.L. and A.G. Clark. 1997. Principles of Population Genetics, 3rd ed. Sinauer, Sunderland, USA.

Lewontin, R.C. 1974. The Genetic Basis of Evolutionary Change. Columbia University Press, New York, USA.

Moczek, A.P. and D.J. Emlen. 1999. Proximate determination of male horn dimorphism in the beetle, *Onthophagus taurus* (Scarabeidae, Coleoptera). Journal of Evolutionary Biology 12 : 27-37.

Moczek, A.P. and D.J. Emlen. 2000. Male horn dimorphism in the scarab beetle, *Onthophagus taurus*: do alternative reproductive tactics favour alternative phenotypes? Animal Behaviour 59: 459-466.

Nevo, E., A. Beiles, Y. Gutterman, N. Storch and D. Kaplan. 1984. Genetic resources of wild cereals in Israel and vicinity: I. Phenotypic variation within and between populations of wild wheat, *Triticum dicoccoides*. Euphytica 33: 717-735.

Roskam, J.C. and P.M. Brakefield. 1966. A comparison of temperature-induced polyphenism in African *Bicyclus* butterflies from a savannah–rainforest ecotone. Evolution 50: 2360-2372.

Scheiner, S.M. 1993. Genetics and evolution of phenotypic plasticity. Annual Review of Ecology and Systematics 24: 35-68.

Sokal, R.R. and F.J. Rohlf. 1995. Biometry. 3rd ed. Freeman, New York, USA.

Spiess, E.B. 1977. Genes in Populations. Wiley, New York, USA.

Via, S. and D.J. Hawthorne. 2005. Genetic correlation, adaptation and speciation. Genetica 123: 147-156.

Waddington, C.H. 1957. The Strategy of the Genes. George Allen & Unwin, London, UK.

Wallace, A.R. 1889. Darwinism. MacMillan, London, UK.

Ward, B.L., W.T. Starmer, J.S. Russell and W.B. Heed. 1974. The correlation of climate and host plant morphology with a geographic gradient of an inversion polymorphism in *Drosophila pachaea*. Evolution 28: 565-575.

Windig, J.J. 1994. Genetic correlation and reaction norms of the tropical butterfly *Bicyclus anayana*. Heredity 73: 459-470.

Wood, R.J. and V. Orel. 2001. Genetic Prehistory in Selective Breeding: a prelude to Mendel. Oxford University Press, Oxford, UK.

Wool, D. and S. Mendlinger. 1973. The eu mutant of the flour beetle, *Tribolium castaneum* Herbst: environmental and genetic effects on penetrance. Genetica 44: 496-504.

Wool, D. 1985. Blind flour beetles: variable eye phenotypes in the microcephalic (*mc*) mutant of *Tribolium castaneum* (Coleoptera: Tenebrionidae). Israel Journal of Entomology 19: 201-210.

Wool, D. and D.F. Hales. 1997. Phenotypic plasticity in Australian cotton aphid (Homoptera: Aphididae): Host plant effects on morphological variation. Annals of the Entomological Society of America 90: 316-328.

Wool, D. and T. Agami. 2000. Response to selection and genetic regulation of esterase activity variation in *Tribolium confusum*. Selection 1-3: 205-215.

CHAPTER 11

Genetic Variation in Natural Populations (continued)

ELECTROPHORETIC VARIATION

Electrophoresis is a technical method of separating protein molecules in solution, by their size and/or electrical charge. The operation is performed in a gel made from a polymer in electrolyte buffer (starch gels or acrylamide gels are most often used).

The application of electrophoresis to the study of genetic variation in populations of *Drosophila* (Hubby and Lewontin, 1966; Lewontin and Hubby, 1966) opened the way to the investigation of hundreds of animal and plant species, which revealed large and previously unsuspected amounts of genetic variation in populations (review in Nevo, 1978; Nevo et al., 1984). Literally thousands of research projects employed electrophoresis over a period of 30 years or so. Although the enthusiasm over electrophoresis has diminished somewhat with the appearance of molecular methods on the scene in the 1990s, it is still being used because it is cheaper, simpler to carry out, does not require sophisticated equipment, and the results are easier to interpret than when using molecular techniques.

Rationale

Protein molecules are composed of chains of amino acids. All amino acids have a skeleton of one or more carbon atoms, an acid (–COOH) terminal, an amino group (–NH$_2$), and a residue (R). The acid and the amino

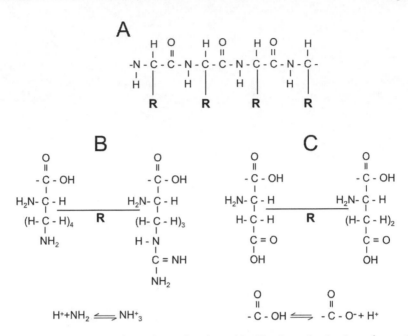

Fig. 11.1 The structure of proteins and amino acids. A) schematic structure of a peptide: C = carbon, H = hydrogen, O = oxygen, N = nitrogen atoms. R = residue. B) 'basic' amino acids. Lysine and arginine have an amino (–NH$_2$) group in R, that dissociates in solution and gives the molecule a positive charge. C) 'acidic' amino acids. Aspartic and glutamic acids have a carboxyl group (–COOH) in R, which dissociates to give the molecule a negative charge.

terminals connect the amino acids to each other to form proteins. The residue is characteristic of each amino acid (Fig. 11.1). The cumulative effects of the residues give the proteins their electrophoretic properties.

In 16 of the 20 common amino acids, the residue is electrically neutral. In lysine and arginine the residue contains an amino terminal, which dissociates in solution to give the residue a negative charge. Glutamic and aspartic acids have has a carboxyl terminal in the residue, which dissociates in solution to give the residue a positive charge. The sum of all charges gives the protein molecule a positive or negative charge, as the case may be (Fig.11.1).

When the gel is placed in an electric field, the proteins migrate towards one of the electrodes. Usually the gel is run in a high pH, and the proteins migrate towards the anode. Their position in the gel is detected with special staining techniques. Many of the studied proteins have enzymatic activity, and are detected by immersion in a specific substrate solution and applying a specific stain. Many staining recipes were published (e.g., Brewer and

Sing, 1970; Shaw and Prasad, 1970). One such gel was illustrated in Fig. 10.9.

The use of electrophoresis in population genetics and evolutionary research is based on the following assumptions (Lewontin, 1974):

(1) Every band (= isozyme) on the gel represents a different protein, the product of a different gene.

(2) A difference in the migration distance between two allozymes (= allelic isozymes) is due to a mutation in the gene coding for that protein.

(3) Proteins migrating to the same distance on the gel are products of the same allele.

(4) There is no dominance in the expression of a gene at the protein level. In a heterozygote, both alleles should be present.

(5) The sampled genes are a random sample of the entire genome.

If a number of samples from different individuals are run side by side, and stained for a specific protein, differences in isozyme migration distance reveal those individuals in which a mutation has occurred in the gene coding for this protein. (If they have enzymatic activity, the variant proteins are referred to as **isozymes**. Proteins resulting from variant alleles at the same locus, are referred to as **allozymes**.) If the individual is a heterozygote, the electrophoretic pattern may show two bands for that particular protein (or more, if the protein is composed of two or more subunits).

Changes in different genes can be identified by using different substrates in the staining recipe. Multiple alleles at a locus may be recognized by different mobilities of the isozymes.

Difficulties

In the thirty years of application of electrophoresis as a tool in population genetics, difficulties were encountered with each of the five assumptions listed above.

Assumption 1: When a small insect (like a *Drosophila* fly or an aphid) is crushed, the 'homogenate' contains not only the proteins of that insect but also proteins belonging to, or produced by, parasites or symbionts of that organism. These can easily be erroneously recorded as genetic variants of the studied organism (examples were reported by Tomiuk and Wöhrmann, 1980; Waterhouse, 1981).

Assumption 2: The migration distance of the isozymes may be affected by non-genetic factors. Migration distances of isozymes of the snail *Cepaea nemoralis* were affected by the food the snails ate (Oxford, 1975, 1978).

Assumption 3 can only be tested by comparison of the DNA sequences of the genes coding for the proteins. This was carried out for only a few isozymes (e.g., ADH isozymes in *Drosophila*).

Assumption 4 is partly a technical matter. Sometimes two bands on the gel are too close together to be detected as separate by the human eye. (A way to bypass this difficulty is to calculate expected heterozygosity with Nei's formula $H = 1 - \Sigma p_i^2$ (see below)).

Assumption 5 is the most difficult. All the isozymes are necessarily structural proteins: regulatory genes are not detectable by electrophoresis. Insoluble proteins or those attached to membranes or other heavy cell particles, are more difficult to separate electrophoretically.

The function of most isozymes in the living body is unknown. The isozymes are identified in vitro by their ability to break down a specific substrate – most often a chemically-defined, artificial substrate (the standard substrate for esterases is α-naphthyl acetate, which does not exist in the living cell). Whole groups of isozymes are analyzed using the same substrate, and separated (and given different names) by the application of different stains. It is not impossible that one broad-substrate enzyme is counted as several specific ones.

Despite all these difficulties, which should be kept in mind, electrophoresis has been a very useful and efficient method for the study of genetic variation in natural populations. The method is straightforward, easy to apply, and large samples can be run in a short time. These characteristics contributed to its popularity and widespread application in the 1960s–1980s.

QUANTITATIVE ESTIMATION OF ELECTROPHORETIC VARIATION IN NATURAL POPULATIONS

Reviews of surveys of genetic variation by electrophoresis have been published (e.g. Nevo, 1978; Nevo et al., 1984). Hundreds of plant and animal species and even strains of bacteria and fungi were surveyed. The following statistics are used as quantitative measures of electrophoretic variation for comparisons among populations or species:

(1) The proportion of polymorphic loci (of the total number surveyed). The definition of a locus as polymorphic is arbitrary and depends on sample size. In most studies a locus is listed as polymorphic if there are two or more detectable alleles and the rarest of them occurs at a frequency of more than 5% (or 1% in other studies). The proportion of polymorphic loci, **P,** varies greatly between populations and species.

(2) The average proportion of heterozygotes per locus, **H**. This statistic may be obtained directly by counting the heterozygotes on the gels, but is estimated most frequently from Nei's formula

$$H_i = 1 - \Sigma p_i^2 \quad \text{for one locus, and}$$

$$\mathbf{H} = \Sigma^n H_i/n$$

where n is the number of loci analyzed. The estimates of heterozygosity are often affected – sometimes strongly – by a large number of invariant (monomorphic) loci. In some samples, 70% of the loci are monomorphic.

There are wide differences in **H** among populations and among species. In general, vertebrates are less variable than invertebrates (in Nevo, 1978: **H** in invertebrates: 0.112, vertebrates 0.049; Nevo's survey included 243 studies in which more than 14 loci were surveyed). A comprehensive review of electrophoretic variation in natural populations was published by Nevo et al. (1984) with emphasis on the ecological implications of this variation.

'NULL' ALLELES

In many studies of natural as well as laboratory populations, individuals were discovered that seemed to lack activity of a particular isozyme, which was active in other individuals of the same population. This property of 'no activity' was shown to be heritable in crosses of such organisms with genetically-marked laboratory strains. In many cases, individuals heterozygous for such 'null alleles' were discovered in samples collected in nature. This phenomenon was particularly common in housefly esterases (Narang et al., 1976) and in *Drosophila* (Roberts and Baker, 1973; Langley et al., 1981).

The phenomenon of 'null' alleles raises some basic biological questions: what makes a normal gene produce a 'no-activity' enzymatic protein? And, how is the life of the individual affected by the inactivity of an enzyme? Houseflies homozygous for the null-allele at the α-glycerophosphate dehydrogenase (α-GPDH) locus, could not fly (the enzyme is active in the flight muscles (O'Brian and Shimada 1974)) – but *Drosophila* flies which lacked any activity of α-GPDH were 'synthesized' by crosses, and could be maintained in the laboratory and used in genetic crosses with field-collected flies (David et al., 1976).

There may be alternative pathways for performing whatever function any enzyme has in the physiology of the animal. If one pathway is blocked by mutation, activity may be enhanced by other pathways (Kacser and Burns, 1980).

MOLECULAR METHODS IN EVOLUTIONARY RESEARCH

Electrophoresis is still being used today (2005) in the study of populations. But the central weight of population genetic studies has been diverted to the use of molecular methods – which enable the estimation of variation at the ultimate, DNA level. (Although electrophoretically detectable proteins are also molecules, the term 'molecular methods' is restricted to those that enable the reading of nucleotide sequences of DNA and RNA.)

In the last two decades of the 20th century, the improvement of molecular methods opened the way to studies of molecular evolution and phylogeny, with ramifications into medicine, police forensic work, classification and taxonomy, ecology and conservation of endangered species.

When two or more DNA sequences are compared, differences between them due to point mutations (and of course larger sequence changes) can be detected. If one sequence comes from an organism – or a gene – which is ancestral to the other (= more primitive), the amount of phylogenetic divergence between the two can be deduced for the construction of phylogenetic trees.

Three methods are most frequently employed: Restriction Fragment Length Polymorphism (RELP) was historically the first to gain widespread use. Mini-satellite analysis ('DNA fingerprinting') is used for particular cases. The most widespread method is Random Amplified Polymorphic DNA (RAPD), where specific sequences of DNA (originally randomly cut, but today constructed specifically by the experimenter and identified by using specific 'primers') are amplified: many copies of the same sequence are produced by Polymerase Chain Reaction (PCR).

The first two methods result in fragments of DNA of variable sizes. The fragments can be separated electrophoretically by their size and detected in agarose gels. PCR (see below) yields sufficient quantities of a specified DNA fragment, which can be 'read' to give the exact sequence of nucleotides. The same fragment can be obtained from different organisms and the DNA sequence can be aligned and compared.

A detailed description of these methods is beyond the scope of the present book. The reader is referred to specific texts like Graur and Li (2000). But a short description may be in order.

RELP

DNA is extracted from tissue homogenates by alcohol precipitation and centrifugation. Then the DNA molecule is cut, using specific **restriction enzymes**. These enzymes are extracted from bacteria and sold commercially in pure form (their commercial names are derived from their bacterial origin, like ECOR1 from the bacterium *Escherichia coli*). These enzymes have well-defined properties. Each enzyme 'recognizes' a specific nucleotide (base) sequence in the DNA molecule and cuts the molecule at any site where the sequence occurs. The result is a mixture of DNA fragments of different lengths, which are separated electrophoretically in agarose gels (small fragments migrate faster). By adding a stain like ethidium bromide, which can be detected in uv light, the pattern of fragment lengths can be visualized and photographed. A 'ladder' (a mixture of DNA sequences of known lengths) is run in parallel for estimation of the fragment lengths in the sample.

This method is particularly useful – and has often been used – in studies of polymorphism in mitochondrial DNA. Mitochondrial DNA is a small, cyclical molecule (~ 16,000 base pairs) carrying a rather limited number of genes which can be compared in different organisms. MtDNA is inherited maternally, and can be used to trace maternal ancestry in phylogeny.

Mini-satellite DNA ("Fingerprinting")

Short sequences of 10-15 base pairs are repeated many (and a variable number of) times next to each other in the genome of many organisms ("mini-satellites"). When the repeated DNA sequence is known, a radioactive probe can be made to attach to it to facilitate its detection and recognition in gels. Since the number of repeats is very large and variable, the probability of finding exactly the same sequence in different individuals is very small, close to zero.

This enables the use of mini-satellites to identify individuals by the DNA pattern, in much the same way as fingerprinting is used to identify individual humans (Jeffrys et al., 1985a, b). The method serves in human paternity disputes in the courts and in police forensic work. Even minute samples of DNA can be amplified using PCR (see below) to provide enough DNA for analysis. The probability of identity between two individuals – using many probes simultaneously – is as low as 3×10^{-11} (Armour and Jeffrys, 1992).

The use of this method is not restricted to human DNA. It has been used in studies of cats and dogs (Jeffrys and Morton, 1987). Cohen and Dearborn (2004) used DNA fingerprinting to measure the genetic similarity between male and female nest-mates in a colony of great frigate birds in Hawaii. The analyses of parents and fledglings in birds like nesting swallows, which seem to be paired faithfully, surprisingly showed that extra-pair matings were very common (Wetton et al., 1992).

THE POLYMERASE CHAIN REACTION (PCR)

This method enables the amplification of minute quantities of a sequence – specified by the investigator – to provide enough material for analysis.

The method is based on the process of DNA replication. When the temperature is elevated – over 90°c – The two helical strands of DNA separate. When the mixture is cooled in the presence of the necessary nucleotides, and the enzyme required for DNA replication, each strand serves as a template for the formation of the complementary strand and the quantity of DNA is thus doubled. The procedure is repeated – usually more than 30 times – in an automatic temperature cycler, which changes the temperature from high to low very quickly.

The key to the success of the method was the isolation of an enzyme, Taq Polymerase, which is now available commercially. It was isolated from bacteria which live in hot-water springs, and its activity is not destroyed in temperatures near the boiling point.

The PCR method is very useful for studies of molecular phylogeny, because even minute samples of DNA from preserved museum specimens or from archeological and palaeontological finds, can provide sufficient material for analysis.

Sequencing machines are now available that can 'read' the sequence of nucleotides in the sample automatically (Lewin, 1983; Graur and Li, 2000). But despite the great advantage and technical improvements, molecular methods are not as widely used in population genetics studies as electrophoresis was in the late 20th century. The methods are still too complex and expensive to be used on many individuals within species. The data banks (which are becoming more numerous every year) contain sequences of many genes but still represent only a handful of species, and molecular phylogenetic analyses are often carried out on single species from each family.

Literature Cited

Armour, A.J. and A.J. Jeffrys. 1992. Biology and applications of human minisatellite loci. Current Opinions in Genetics and Developent 2: 850-856.

Brewer, G.J. and C.F. Sing. 1970. An Introduction to Isozyme Techniques. Academic Press. New York, USA.

Cohen, L.B. and D.C. Dearborn. 2004. Great frigatebirds, *Fregata minor*, choose mates that are genetically similar. Animal Behavior 68: 1229-1236.

David, J.R., C. Bouquet, M.F. Arens and P. Fouillet. 1976. Biological role of alcohol dehydrogenase in tolerance of *Drosophila melanogaster* to aliphatic alcohols: utilization of an ADH-null mutant. Biochemical Genetics 14: 989-997.

Graur, D. and W-H. Li. 2000. Fundamentals of Molecular Evolution. Sinauer, Sunderland, USA.

Hubby, J.L. and R.C. Lewontin. 1966. A molecular approach to the study of genic heterozygosity in natural populations. I. The number of alleles at different loci in *Drosophila pseudoobscura*. Genetics 54: 557-594.

Jeffrys, A.J., W. Wilson and S.L. Thein. 1985a. Individual-specific "fingerprinting" of human DNA. Nature 316: 76-79.

Jeffrys, A.J., W. Wilson and S.L. Thein. 1985b. Hypervariable "minisatellite" regions in human DNA. Nature 314: 67-73.

Jeffrys, A.J. and D.B. Morton. 1987. Fingerprinting in dogs and cats. Animal Genetics 18: 1-15.

Kacser, H. and J.A. Burns. 1980. The molecular basis of dominance. Genetics 99: 639-666.

Langley, C.H., R.A. Voelker, A.J. Leigh Brown, S. Ohnishi, B. Dickson and E. Montgomery. 1981. Null allele frequency at allozyme loci in natural populations of *Drosophila melanogaster*. Genetics 99: 151-156.

Lewin, B. 1983. Genes. Wiley, New York, USA.

Lewontin, R.C. 1974. The Genetic Basis of Evolutionary Change. Columbia University Press, New York, USA.

Lewontin, R.C. and J.L. Hubby. 1966. A molecular approach to the study of genic heterozygosity in natural populations. II. Amount of variation and degree of heterozygosity in natural populations of *Drosophila pseudoobscura*. Genetics 54: 595-609.

Narang, S., A.C. Terranova, I.C. McDonald and R.A. Leopold. 1976. Esterases in the housefly: Polymorphism and inheritance patterns. Journal of Heredity 67: 30-38.

Nevo, E. 1978. Genetic variation in natural populations: pattern and theory. Theoretical Population Biology 13: 121–177.

Nevo, E., A. Beiles and R. Ben Shlomo. 1984. The evolutionary significance of genetic diversity: ecological, demographic, and life history correlates. In: G.S. Mani (ed.), Evolutionary dynamics of genetic diversity – lecture notes in biomathematics 53: 13-213.

O'Brien, S.J. and Y. Shimada. 1974. α-glycerophosphate cycle in *Drosophila melanogaster*. IV. Metabolic, ultrastructural and adaptive consequences of α-GPDH-1 "null" mutations. Journal of Cell Biology 63: 864-882.

Oxford, G.S. 1975. Food induced esterase phenocopies in the snail *Cepaea nemoralis*. Heredity 35: 361-370.

Oxford, G.S. 1978. The nature and distribution of food induced esterases in helicid snails. Malacologia 17: 331-339.

Roberts, R.M. and W.K. Baker. 1973. Frequency distribution and linkage disequilibrium of active and "null" esterase isozymes in natural populations of *Drosophila montana*. American Naturalist 107: 709-726.

Shaw, C.R. and R. Prasad. 1970. Starch gel electrophoresis of enzymes: a compilation of recipes. Biochemical Genetics 8: 79-119.

Tomiuk, K.J. and K. Wöhrmann. 1980. Enzyme variability in populations of aphids. Theoretical and Applied Genetics 57: 125-127.

Waterhouse, W.J. 1981. An esterase of bacterial origin in *Drosophila montana*. Biochemical Genetics 19: 227-231.

Wetton, J.H., D.T. Parkin and R.E. Carter. 1992. The use of genetic markers for parentage analysis in *Passer domesticus* (house sparrow). Heredity 69: 243-254.

CHAPTER 12

Evolutionary Processes in Natural Populations

ECOLOGICAL GENETICS

There is no doubt that natural populations are genetically variable. But evidence is still required that natural selection uses this variation to modify natural populations and that new species evolve in the way that Darwin anticipated.

Natural selection is expected to adapt populations to their environments. The selective forces are ecological. Environmental changes may drive genetic changes in populations, as demonstrated by the widespread insecticide resistance following the application of insecticides for pest control. Environmental changes induced by the activity of the organisms themselves may affect selective processes: ants and mole rats digging nests and tunnels underground, create microenvironments for other organisms; and gall-inducing insects modify the plant tissues in a manner favorable to themselves (e.g., Wool et al., 1999), creating protected environments for their progeny (and associated parasites, predators and inquilines). These changes may affect the future evolution of the associated organisms.

The interaction of the population with its environment determines both the direction and magnitude of the evolutionary change. The branch of evolutionary science dealing with these interactions has been termed Ecological Genetics. Its founders were E.B. Ford and his students in England (e.g., Ford, 1964; Creed, 1971), who designed and carried out many field trials in order to measure these interactions and to demonstrate that natural selection actually brings about genetic changes and adaptation in populations.

Despite the efforts invested in such investigations, and the importance of the issues for the science of evolution, only a handful of cases have been published to date in which selective processes in nature were followed in detail. These few cases may serve as models, however, because they increase our confidence that natural selection is the best explanation for the majority of observed natural phenomena: generally one can only observe the end results of long-term evolutionary processes.

In a literature survey, Reznick and Ghalambor (2001) found 47 cases where ecological pressures caused rapid adaptation of natural populations in the last 200 years. In some of these studies the authors provided evidence that a genetic change had occurred. Most of these cases involved a response to an anthropogenic disturbance of the habitat, such as environmental pollution (leading to industrial melanism, see below) or the use of pesticides. The accumulation of toxic soils at the 'tailings' of mines enabled the evolution of heavy-metal tolerance in plants. In other cases, evolutionary changes were noted in body sizes of fish and birds, morphological and physiological characters in invertebrates and others. Almost all the reported cases (41 of 47) involved colonization of new sites or shifting to a new host plant.

Reznick and Ghalambor (2001) suggest that the reason that so few cases of contemporary evolution have been reported may be not that the processes in nature are too slow, but on the contrary – that they are too fast to be detected: unless one happens to be studying the right character at the right time and place, the process may be missed!

INDUSTRIAL MELANISM IN MOTHS

One of the first to be described, and perhaps still the most famous, example of genetic changes in populations of organisms following a change in the environment, is the phenomenon called industrial melanism. It attracted public attention (not only in scientific circles) in the early 20th century. This was because that particular environmental change was very obvious and affected humans, not only the insects that gave the phenomenon its evolutionarily importance.

In the middle of the 19th century, dark or black individual moths, belonging to otherwise white or mottled species, began to be noticed by collectors in England (Fig. 12.1). The dark morphs were first noticed in the industrial districts in central and south-eastern England, and this association with industry gave the phenomenon its name.

Fig. 12.1 Melanic and typical morphs of *Biston betularia* (Geometridae) resting on a lichen-covered (unpolluted) tree bark (left) and on a polluted, blackened bark (right) (from Ford 1964). The melanic moth is very conspicuous on the lichen-covered bark, but blends with the blackened background. The typical moth is inconspicuous on the lichen-covered bark.

In the early 20th century the phenomenon was widespread enough to attract scientific attention. By then, the frequency of dark or black ('melanic') moths had reached 95% in some areas, and the number of species with melanic moths approached 80, not only in England but also in industrial areas in other countries in Europe.

A common behavioral characteristic of these species – which were not always related taxonomically – was that they were largely nocturnal or active during dusk hours, and spent the day resting on trunks and branches of trees with their wings spread, as many moths do. The wings of the normal ('typical') forms of these moths were light gray or white, with spots or thin line marking.

It was suspected that the appearance of the dark moths was somehow related to environmental pollution (Ford, 1964; Brakefield, 1987; Berry, 1990). Inspection of moth collections in the British Museum – and in privately-owned collections – revealed that black moths were very rare in the mid-19th century, but fifty years later they were the majority – close to 100% in some areas. Those fifty years had seen the great environmental impact of the Industrial Revolution in Europe. The smoke and soot from the chimneys of mines and factories – and private homes where coal was used for heating – visibly affected the landscape. One of the most noticeable changes (to scientists at least) was that lichens, which normally covered the trunks and branches of trees in the wet forests of England, disappeared, and the bare trunks and leaves were covered in black soot. Experiments in washing the leaves showed that the soot was blown by the winds far from the industrial areas. The smog in London became infamous.

In the 1930s, a number of scientists suggested that some poisonous chemicals in the soot had caused melanic mutations to accumulate in the moths (these were the years following the rediscovery in 1900 of Mendel's paper on the inheritance of color in garden peas, genetics was emerging as a science, and mutations were popular. See review in Huxley, 1942). But experiments disproved this suggestion: the dark morph was not induced by poisonous mutagens in the polluted air (Ford, 1964).

A plausible explanation, based on natural selection, was offered by Ford, and his students – notably Kettlewell – set out to examine that explanation in field trials (Ford and his followers emphasized the need for carefully controlled experiments and observations in nature. This was a novelty at the time: discussions about evolution in the 19th and early 20th centuries were theoretical, 'armchair biology').

Kettlewell's Field Experiments

Kettlewell chose to work mainly with one species, the peppered moth *Biston betularia* (Geometridae*)*. This moth has a typical morph, with light, spotted ('peppered') wings, and a dark, melanic form, called <u>carbonaria</u>. Kettlewell's work was publicized in popular articles (e.g., Scientific American, 1953), books (e.g., Kettlewell, 1973) and movies and became a textbook example of the operation of natural selection.

Genetic crosses indicated that the dark morph of *B. betularia* was due to a dominant allele (confirmed by Mikkola, 1984). Why was this morph so rare before 1850? Ford's suggested that natural selection worked against the dark morph. Its frequencies were kept low because predators, searching visibly for moths, noticed it easily on the lichen-covered tree trunks and branches, where the typical morph was inconspicuous. When the environment changed, the situation was reversed: the light-colored typicals were much more conspicuous on the blackened tree trunks (Fig.12.1), and were selected against by the predators.

This hypothesis was tested by Kettlewell in three field experiments – one in a polluted site near Birmingham, in 1953, and two in a non-polluted site in Dorset – two years later – one of which was a repeat of the 1953 study (Hagen, 1999).

The methods were similar in all experiments. Several hundred laboratory-reared moths (marked by a dot of paint on the wings) in approximately equal frequencies of the two morphs, were released in a clearing in the forest. Light traps, augmented by cages baited with live unmated females, were placed in the clearing to attract the moths. Recaptured moths were separated by color and counted. In the latter two experiments, the moths were released in the evening. The following day, researchers stood in the clearing armed with binoculars and cameras and documented and photographed the moths which were attacked or captured by birds.

In the first study (1953) in the polluted environment, twice as many of the released melanic moths were recaptured in the traps as the typicals. This was interpreted as showing that the light-colored typicals were selectively removed by the birds, while the melanics escaped predation in the polluted environment.

The other two experiments were run in parallel. In an unpolluted area in Dorset, the results were reversed: twice as many typical than melanic moths were recaptured in the traps. Many more melanic than typical moths were captured by birds: of 190 moths photographed, 26 were typical and 164 were melanics (Ford, 1964). Predators were mainly thrushes (*Turdus*), robins and flycatchers.

Repeating the 1953 experiment in the polluted area in Birmingham, most of the recaptured moths were melanics (tables in Hagen, 1999).

These results illustrated an evolutionary process 'in action'. In all experiments, the camouflaged morph was better protected from predators. The birds were changing the genetic composition of the population by selectively removing the less adapted morph.

Ford suggested that the rapid change in frequency of the dark morph in the 19th century was due to a balance between two factors. The larvae of the dark morph probably had some physiological advantage, but in the unpolluted forest the melanic adults were selected against by bird predation. Elimination of the lichens and darkening of the environment when the soot covered the trees, made the melanics relatively better protected than the typicals, enhancing their physiological advantage (J.B.S. Haldane calculated in 1924 that the dark morph probably had a 30% advantage over the typical).

Industrial melanism illustrates that natural selection can change its course with a change in environmental conditions: a formerly protected morph lost its protection, while a morph formerly exposed to natural selection became protected.

More Recent Studies

The amount of work invested in studies of industrial melanism attests to its importance as the central example of natural selection 'in action' (Kettlewell and Berry, 1961; Kettlewell, 1973; Lees et al., 1973; Cook et al., 1990; Brakefield, 1987; Berry, 1990). A confirmation of the relation of melanism to environmental pollution was obtained in a long-term follow-up study of *Biston betularia* (Clarke et al., 1985).

Kettlewell's studies raise a number of questions: 1) Is bird predation strong enough to account for the variation in frequency of dark morphs in different polluted areas? 2) If not, what other factor(s) should be considered to explain this variation? 3) How general is this mechanism – are there other species, and organisms other than moths, where melanics show an advantage in polluted environments? 4) In these other cases, is bird predation involved?

Geographic variation in the frequency of melanic moths has been observed since Kettlewell's early work. Detailed research has been aimed at quantifying pollution and testing for the efficacy of bird predation as the explanation for the geographical variation of melanic frequency. Traps for moths were set along a transect from the city of Liverpool to the rural countryside (Bishop, 1972). The frequency of trapped melanics in fact decreased from about 85-95% near the city to 5-10% in the rural areas in Wales, 50 km away.

The effect of predation was assessed by placing fresh-frozen typical and melanic moths on eight trees in each of seven woods. The moths still remaining on the trees were counted daily and those lost were replaced. Fewer melanic than typical moths were removed in the polluted areas near the city, but quantitatively the difference was too small to fully explain the observed geographical change in melanic frequency between Liverpool and rural Wales (Mani, 1980).

Other factors were tested: for example, the effect of wetting the tree trunks (wet bark looks darker than dry bark) (Lees and Creed, 1975). In a computer simulation, estimated differences in moth longevity, and flight distance of the moths per night, were added to predation. A combination of estimated values of all these parameters improved the agreement of the simulation results with the observed geographical pattern.

Another direction of research involved the analysis of the components of polluted air and their effect on the disappearance of the lichens from the tree trunks (lichens are very sensitive to air pollution and are used as bio-indicators for it). Endler (1984) measured quantitatively the similarity between moth wing pattern and the lichen-covered background. He found that the moths were best camouflaged against the most common lichen phenotype. The disappearance of lichens was the first indication of pollution, and both the color and the texture of the substrate on which the moths were resting changed when the lichens were destroyed. Lees et al. (1973) measured six variables of vegetable cover and 10 other physical variables on 10 oak trees in each of 104 sites in England to evaluate their impact on the percentage of melanic *B. betularia*. The variable most closely correlated with melanic frequency was the content of sulphur dioxide (SO_2^-) in the air, and not the amount of carbon particles (soot), which contributed most to the blackened appearance of the landscape.

A search for other contributing factors to the frequency of melanics led to a re-examination of moth behavior. Mikkola (1979, 1984) claimed that the field work with frozen moths was based on a wrong assumption: most of the moths did not rest on tree trunks at all – they tended to rest on thin, lichen-covered branches at the top of the canopy, where they were least exposed to selective predation (Fig. 12.2)

Mikkola suggested that predation may be effective when the moths emerge from their pupae in the leaf litter under the trees. Very early in the morning, newly-hatched moths climb on tree trunks to expand their wings,

Fig. 12.2 *B. betularia* camouflaged when resting on a thin lichen-covered branch (from Mikkola, 1979; reprinted with permission of Annales Entomologici Fennici).

and selection may take place at this stage. Liebert and Brakefield (1987) found that pairs of mating moths in copula are conspicuous even if only one of the pair is melanic.

In 1956, a strict 'clean air' law went into effect in England. It banned the use of coal for heating houses and factories and replaced it with oil products, which produce far less smoke (it is historically interesting that the reasons for enacting the law were not necessarily the interest in public health or scientific principles. The direct reason was the death of some valuable cows in an agricultural exhibition and the canceling of opera performances as the audience could not see the stage because of the smog. Berry, 1990, p. 308). As a result of the law, air pollution was greatly reduced, and the rain gradually washed away the soot from the leaves. Surveys in England showed a gradual but consistent trend of decrease in the frequency of melanic moths (Cook et al., 1986; Clarke et al., 1985).

Interestingly, a moth in the USA was identified as *B. betularia* (perhaps a different subspecies). In Michigan, a rare melanic variant (1-3% of the population) was discovered (West, 1977). In an industrial area in south Michigan, the melanic constituted about 90% of the moths! The population was not sampled for many years. When sampling was resumed in 1994-1995, melanic frequencies were as low as in England. There was no change in lichen cover on the trees, but the level of sulphur dioxide in the air had been drastically reduced (Grant et al., 1996). The similarity of the trends on the two sides of the Atlantic remains unexplained.

SHELL COLOR POLYMORPHISM IN THE EUROPEAN LAND SNAIL

The genetics of the European land snail, *Cepaea nemoralis*, and its congener *C. arvensis*, was intensively studied in the 1950s. The phenotype of the snail's shell color is controlled by two sets of genes – one set determines the background color (brown, pink or yellow) and the other, the number of bands on the shell (0 to 5). These phenotypic traits can be observed on both live and dead snails and even on broken shell fragments (Cain and Sheppard, in Ford, 1964).

The snails are consumed by birds. In the area investigated by Sheppard, the main bird predators were thrushes (*Turdus ericetorum*). To break the shells, the birds smash the snail on some available rock in their territory. The broken shells remain scattered around the rock (bird 'anvil' is the term used by Cain and Sheppard).

The investigators compared the phenotype frequencies in broken shells near the bird anvils – with random samples collected in the same area. The results showed that the birds were not sampling the snails at random: they were taking the most conspicuous phenotypes more often than their frequencies in the field. When the yellow winter color of the grass changed to green in spring, there was a change in the proportion of yellow shells around the anvils: In early spring 42% of the snail shells sampled near the anvils, but only 24% of the randomly sampled snails, were yellow. In later samples, the proportions in the random field samples did not change, but the proportion of yellow snails at the anvils was reduced to only 10-15% of the total.

As in the case of industrial melanism, this too seemed to be a case of natural selection in action, with predation by birds causing a seasonal shift in the frequencies of differently-colored snails.

The reports of Cain and Sheppard prompted other scientists to work on the ecological genetic processes in populations of the snails. But 45 years of intensive research, which yielded more than 300 scientific publications, failed to demonstrate that selective predation plays a major role in the differentiation of color patterns in snail populations (reviews in Jones et al., 1977; Goodhart, 1987). In the 1970s, most of the research was aimed at explaining the geographical variation in snail polymorphism by climatic variation and 'founder effects' followed by local population expansion. The observed mosaic of patches with locally different proportions of the phenotypes was referred to as 'area effects' in the literature (Cameron and Pannett, 1985; Cameron, 1992).

No relationship was found between characteristics of the habitat and the frequency of electrophoretic markers in the snail populations. No reproductive barriers between snails from different populations were found in the laboratory, although the snails were sampled from different countries and were electrophoretically distinct (Johnson et al. 1984). The simple explanation of polymorphism maintained by bird predation does not seem adequate for the snail populations.

More recent studies

The interest in land snail polymorphism as models for ecological genetic research still persists. The limited dispersal ability of snails suggests that their distribution may form a meta-population. Schweiger et al. (2004) used selectively-neutral molecular markers (microsatellites) as well as morphological banding phenotypes in a study of snail population structure. The authors combined Wright's F-statistics (Chapter 6) and spatial correlation analysis to establish that *C. nemoralis* exists genetically as a network of spatially-interconnected subpopulations. The snail does not colonize arable, agricultural fields, which form a matrix of uninhabited land (by the snail) around suitable habitats.

The mean distances among colonized patches were an order of magnitude larger than the mean annual dispersal distances of the snails. The distribution of dispersal distances was leptokurtic: most individuals moved only short distances in a year.

Schweiger et al. (2004) suggest the following scenario for the establishment of 'area effects': Small numbers of random founders, reaching empty sites, establish unique allele frequencies there (genetic drift). After each population is founded, it expands its distribution by short-distance migration.

INSECTICIDE RESISTANCE

Some cases of insect resistance to insecticides were known before the 1940s, and Dobzhansky described the phenomenon in 1937 as one of the best examples of rapid evolution. Today the resistance of insect pests, or disease vectors, to different insecticides is one of the most important and widespread phenomena in modern agriculture and epidemiology.

Insecticides attracted attention when DDT – which was introduced in the 1940s for the control of lice, vectors of typhoid fever in military and refugee camps during and after World-War II – found widespread use in agriculture and raised high hopes as a weapon against insects. DDT was followed by other, more potent, poisons like organophosphates. After a period of use, however, the insects became resistant to the new poisons – and the race between chemists and insects has been going on since then.

Huxley (1942) lists some early examples of the use of extremely lethal poisons, which are banned today due to their dangerous effects on humans. In California, cyanide gas (HCN) was used to control scale insects in citrus orchards (whole trees were covered in a tent and the gas released there). It was reported in 1914 that the treatment, which had formerly been effective for 2-5 years, no longer killed the pest even when applied twice a year. One year later resistance was reported in the olive scale: the concentration needed to kill the resistant insects was 4 times higher than the dose that caused damage to the trees. Various compounds of arsenic, which were also used, were later banned because they were toxic to humans. Laboratory studies showed already in the 1940s that resistance is hereditary, and that the scale insects were resistant even to chemicals to which they had not been exposed.

Cases of rapid evolution of insecticide resistance are listed by Forgash (1984). Resistance to many new insecticides was reported within two or three years of their first field application. For example, in Egypt, 19 different chemicals were used to control the caterpillars of the moth *Spodoptera littoralis* (a pest of cotton and many other crops). Between 1961 and 1984, the pest became resistant to all 19 chemicals, and no chemical remained effective for more than 2-4 years.

Thousands of research reports have been published on resistance of insects of agricultural or medical importance. Most of these deal with only a few groups of insects – mosquitoes (e.g., Pasteur and Raymond, 1996), houseflies (mainly in Europe), the sheep blowfly (*Lucilia cuprina*) in Australia (e.g., McKenzie et al., 1992), aphids and whiteflies (in the USA and England) and flour beetles (in southern USA).

Only a fraction of these reports deal with the evolutionary aspects of resistance. The assumption is that the resistance-conferring alleles were present, at very low frequencies, in the insect populations before the first application of the pesticide. When the pesticide was applied, susceptible insects were killed, but those which were endowed with the ability to survive in the presence of the new environmental factor became the parents of the next generations of the insects. This is why resistance is considered a model for rapid evolution by natural selection. (A distinction need perhaps be made: the pesticide is man-made, but once introduced into the insect habitat, the pesticide becomes a part of the natural environment to which the population adapts by natural selection.)

In the absence of the pesticide, resistant individuals were probably inferior in fitness to the susceptible ones: this assumption is supported by data that show that if application is suspended for a long enough time, the population may become susceptible again. (Roush and Daly (1990) claim, alternatively, that the population in such cases may become less resistant due to migration of susceptible individuals from untreated areas.)

In many cases resistance is specific to a certain chemical – or group of chemically related pesticides with a similar mode of action (e.g., Mckenzie, 1984). Even if resistance is initially a local phenomenon – which may be the rule – it may spread quickly by migration of resistant insects, assisted in a large degree by human means of transport (e.g., in mosquitoes, Raymond et al., 1991).

It is unclear how many genes are involved in conferring resistance. A number of investigators suggested that resistance may be a polygenic trait (e.g., Firko and Hayes, 1990). In an extensive review, most field cases of resistance seemed to fit a model of a single locus with two or more alleles, while resistance appeared polygenic mostly in laboratory selection experiments (Roush and Mckenzie, 1987; Roush and Daly, 1990). The authors suggested that alleles conferring high resistance are very rare and unlikely to be found in a laboratory population. Any increase of resistance by selection in the laboratory is therefore likely to be the cumulative effect of alleles at several loci. In contrast, although drastic pesticide treatment may kill most of the insects, no treatment kills 100% in the field. Rare, extremely resistant individuals may survive to be the founders of the next generation, and are capable of restoring pest population sizes to the former levels within a short time (Macnair, 1991).

Economic Implications

Insecticide resistance is a top priority, global economic problem which causes great losses of food production. Already in 1975, 305 species of agricultural pests in the USA had become resistant to at least one pesticide. The annual cost of insect control runs to hundreds of millions of dollars.

Pesticides were introduced in the hope of reducing the pest populations. The success with DDT even raised hopes that some pests could be eliminated. But this goal was not reached with any pest. Present methods of pest control aim not to eliminate the pest (with it, a whole assemblage of parasites and predators may disappear – a highly undesirable result), but instead to keep the pest population small enough and below the level of economic damage.

Tests for Resistance

Not all control failures in the field are due to resistance. If control fails, samples of the pest are collected to find out whether resistance was the cause.

There are two general approaches to testing for resistance. In both approaches, mortality of field samples at fixed doses of the pesticide is compared with that of samples from a susceptible laboratory strain, which is used as a standard.

One method is to expose the insects to a substrate sprayed or dipped in increasing concentrations of the pesticide (or, alternatively, keeping the concentration fixed but varying the duration of exposure). This method has the advantage that it simulates the field application, but the disadvantage that there is no control over the exact dose that each insect receives: the dose may depend on the distribution of the spray in the environment, on the behavior of the insect, the quality of the insect cuticle, and other unknown factors.

The second approach overcomes these disadvantages, but is very unlike the field conditions and can be carried out only in the laboratory. A droplet of the pesticide solution, at a known dose and droplet volume, is applied to a specific spot on the individual insect cuticle. This method, called topical application, is suitable for accurate physiological or genetic studies where the exact amount given to each individual needs be accurately controlled.

The statistic used for comparison of the field with the control strains is most commonly LD_{50} (or LC_{50}), the dose (or concentration) that would kill 50% of the individuals. This value is estimated from the regression of percentage mortality on pesticide dose or exposure duration (with an appropriate transformation for normalizing the percentages (See a book on statistics, e.g., Sokal and Rohlf, 1995) (Fig.12.3).

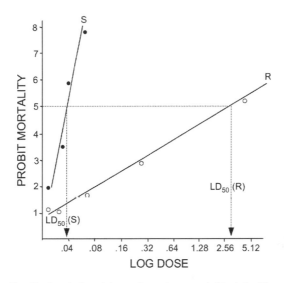

Fig. 12.3 Estimating the level of resistance from dose-mortality data. The abscissa (dose) is usually drawn in a logarithmic scale. Percentage mortality (ordinate) is transformed to probits for statistical reasons.

(Local) Evolution of Insecticide Resistance

The pesticide does not only act ecologically to reduce pest population size, but also as a selective agent changing the composition of the population.

Laboratory investigations with *Drosophila*, houseflies, and a number of species of mosquitoes showed that resistance is hereditary. Alleles conferring resistance may be dominant or recessive, and additional 'modifier' genes are sometimes involved in the phenotypic expression of resistance. But the genetic factors are not the only determinants of the fate of the individual when exposed to the poison (i.e., whether it will live or die). The physiological state, individual weight, and sensitivity to external conditions (e.g., temperature) vary among individuals and influence their fate. The multiplicity of factors affecting mortality may give the false impression of a continuous quantitative trait.

Physiologically, resistance may result from enzymatic processes in the insect body, which neutralize or break down the chemical to harmless components. In many cases, 'mixed-function oxidases' neutralize DDT and PCB-related compounds in mammals. Esterases (carboxyl-esterases) and phosphatases break down organophosphorous insecticides (Perry and Agosin, 1974). The involvement of esterases in the breakdown of organophosphorous insecticides was investigated in great detail in mosquitoes (Mouches et al., 1986; Pasteur and Raymond, 1996), and in aphids (Field et al., 1988; Devonshire, 1989). In resistant individuals of the aphid *Myzus persicae,* the quantity of esterases was several-fold greater than in susceptible individuals. The production of many enzyme molecules is due to the amplification of the gene coding for the resistant esterases (Devonshire and Field, 1991).

The genetic and physiological mechanisms conferring resistance to a pesticide may not be the same even in strains of the same species. In strains of the flour beetle *Tribolium castaneum*, at least two mechanisms of malathion resistance were reported – in one strain, originally collected in Australia, resistance was polygenic (Wool and Manheim, 1980; Wool et al., 1982). In another strain, originally collected in Nigeria, resistance was due to a single, dominant allele (Wool and Noiman, 1983).

'Ecological resistance'

Control failure may ensue from selection for ecological or behavioral traits of the insects, independently of the physiological mechanisms discussed above. Female *Anopheles* mosquitoes (malarial vectors) became resistant to DDT by changing their behavior: instead of resting on the insecticide-treated walls of the dwellings – they alighted directly on the skin of the sleeping humans (Brown 1960, citing earlier work by Trapido). Pinniger (1974) reported that frequent malathion sprays failed to control flour beetles (*Tribolium castaneum*) in a warehouse. When samples of the beetles were tested by topical application, they were found to be as susceptible as the control. Behavioral experiments confirmed that the warehouse beetles were less active and were therefore less exposed to the sprayed surface.

Forgash (1984) reported on great damage to apples from the apple moth in Nebraska, USA: the mature larvae bored their way into the fruit without swallowing the sprayed apple peel. Cockroaches (*Blatella germanica*) from sites where glucose bait was mixed with the pesticide, evolved an aversion to glucose (upon replacing it with fructose, control was resumed). The ability to avoid glucose was heritable (Silverman and Ross, 1994).

Ecological resistance may be more difficult to deal with than physiological resistance (burrowing inside the fruit protects the insect from any kind of chemical that is applied to the apple peel).

To avoid the development of resistance, farmers are advised to frequently change the insecticides and use chemically-different compounds each time. This was shown to be effective in a cage experiment with houseflies (Pimentel and Bellotti, 1976). Caged flies were fed solutions containing six deleterious (but not lethal) chemicals. When each of these compounds was given separately, there was an initial drop in productivity and survival, but within 16 generations the population had become resistant to the chemical, recovered, and increased in size to control level. When all six chemicals were given simultaneously in one cage (pipettes containing them were arranged randomly), the population remained small for double the time (32 generations).

The rapid spread of resistance to antibiotic medicines in bacteria is a specially difficult problem. Due to their rapid reproductive rate (a generation in less than 30 minutes, depending on conditions) and the asexual mode of reproduction, a single surviving resistant bacterium can quickly give rise to a large population and bring back the symptoms of the disease. Therefore bacterial infections are treated with very high doses of antibiotic, in the hope of reaching 100% kill.

HEAVY METAL TOLERANCE IN PLANTS

Plants are expected to evolve more slowly than animals, because their generation times are often longer than in most animals. But research in England showed that evolutionary changes in plant populations may occur relatively quickly. Plants succeeded in colonizing the heaps of waste soil (called <u>tailing</u>) at the mouths of abandoned copper and zinc mines (in Creed, 1971). The soil in these tailings was excavated with the ore from the depth of the mine, and discarded when the ore was sifted out. Research showed that the soil often contained concentrations of the heavy metals 1000 times greater than the normally-lethal concentrations for many plants (Bradshaw, 1971). Plants usually fail to grow roots in solutions containing heavy metals even in minute concentrations (5 ppm of copper is considered lethal for many plants). Plants from the tailings were able to grow in solutions with much greater concentrations of heavy metals than plants from neighboring fields (Bradshaw, 1971). (Fig. 12.4).

When a mine was abandoned, plant seeds from the neighboring fields were probably blown onto the tailings, and some of them took root there. The exact date of the opening and abandonment of these mines are on record. Resistance developed over no more than 80 years. The level of

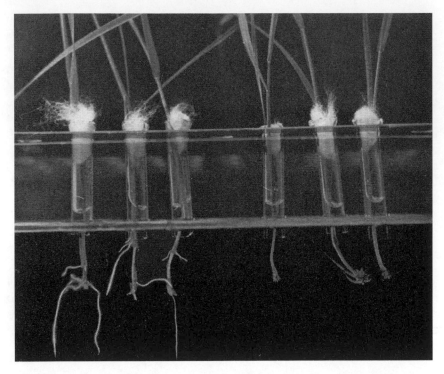

Fig. 12.4 Test for copper tolerance in the grass *Agrostis tenuis*. Seedlings from copper-mine tailings (left) and neighboring fields, grown in a solution containing 0.5 ppm copper (from Bradshaw, in Creed 1971; Reprinted with permission of the publisher).

resistance (e.g. to copper) in samples of field plants decreased in the direction of the prevailing winds – indicating that dust blown from the tailings contaminated the soil in the neighborhood of the mines.

Wu et al. (1975) describe in detail a case of evolution of copper resistance in grasses. A new metal refinery plant was built in the year 1900, and caused heavy air and soil pollution. The dust from the chimneys contained heavy metals like copper, zinc and lead. The concentration of copper in the soil near the plant reached 4000 ppm.

A grass lawn was initially planted in front of the plant, and 70 years later it had a homogenous and neat appearance. However, attempts to enlarge the lawn by sowing seeds collected from nearby fields were unsuccessful. Research showed that the grass from the old lawn was much more resistant to the heavy metals than the grass from neighboring fields (Wu et al., 1975).

The adaptation of the grass populations to contaminated soils is supposed to have started by small numbers of seeds blown by the wind into the tailings. If this scenario is correct, the genetic variation of the colonizing plants should be smaller than in the source field. Vekemans and Lefebvre (1997) tested this idea with electrophoretic markers in the plant *Armeria maritima* (Plumbaginaceae), which grows in soils containing zinc, lead and copper in mining areas of western Europe. Only 6.4% of the variance in the frequency of the markers was due to differences between populations on contaminated and contaminant-free habitats (33% was due to differences between populations within habitats, and 60% due to individual differences within populations). Colonization in this case was not accompanied by a genetic bottleneck.

EXPERIMENTS WITH NATURAL POPULATIONS OF GUPPIES

Commercial fisheries have expressed concern at the observed decline in yield in recent years. The decline may not be due only to depleted stocks: the over-exploited fish populations may be changing genetically (evolving) towards smaller body sizes and lower growth rates, and therefore do not rebound when the selection pressures are reduced.

Reznick and Ghalambor (2005a) review a series of experiments, conducted over a number of years, on natural populations of fish (guppies, *Poecilia*) in mountain streams in the island of Trinidad. Fishing is an exaggerated human analogue of predation; Guppies were used as models: manipulation of predation pressure in natural populations of guppies may simulate the effect of fishing on commercial species.

Guppies have several advantages for research. They are small and have a short generation time. They can be maintained and bred in aquaria at ambient temperatures and the inheritance of their characters can be studied. The males are colorful and variable. The (larger) females produce live progeny so that reproductive rate can be readily measured.

The mountain streams in Trinidad provided very suitable settings for this research. They flow rapidly through waterfalls, which are effective barriers between upper (before the waterfall) and lower pools, where the fish live and reproduce. The upper pools maintain only one species of potential fish predator, while the lower pools have a few cichlid predatory fish which feed on the guppies. Hence predation pressure is naturally higher in lower than in upper pools within a single stream.

The investigators translocated guppies from high- to low-predation pools, as well as translocated predators to low-predation pools, and measured the effect of the change on the guppies (in different experiments)

4, 7, and 11 years after the translocation. Different streams served as replications (Reznick and Ghalambor, 2005a). Among the measured variables were the numbers of offspring per litter, male size, embryo weight and time to maturity.

The results show that guppies from high-predation treatments evolved to smaller sizes and weights, and showed a tendency for early maturity. The females produced litters more frequently and had smaller progeny, compared with fish from low-predation ponds. Some of these effects persisted for 2-3 generations. The conclusion was that predation pressure shaped the guppies life-history characters, and did indeed produce effects similar to commercial fisheries (the results were placed in a wider context by Reznick and Ghalambor, 2005b).

The conclusion seems less definite however when the ecological conditions in the high-and low- predation streams were compared (Reznick et al., 2001). It turned out that high-predation pools were wider and the surrounding forest had more open canopies, thus more sunlight, and consequently higher resource availability (more primary production and more invertebrates) for the fish than low-predation pools. The density of guppies in high-predation ponds was more than twice that of low-predation ponds.Therefore predation may not have been the direct cause of the observed changes in the guppy populations.

RECENT EVOLUTION IN DARWIN'S FINCHES

When in former times an immigrant settled on any one or more of the islands... it would undoubtedly be exposed to different conditions of life in the different islands, for it would have to compete with different sets of organisms... If then it varied, natural selection would probably favour different varieties in the different islands (Darwin, Origin of Species, p. 311).

One hundred years after Darwin's visit, attention was again focused on the finches named after him, in their native islands. Lack (1947) observed that most of the species share the same habitats, nest sites and breeding season, and some species occur on more than one island. Yet no cases of interbreeding were observed (later research with marked birds showed that about 5% of all matings were interspecific, and that interspecific hybrids were fertile. Grant and Grant 1994). Lack claimed that Darwin's notion of 'adaptive radiation' from a common ancestor (Fig. 9.1) was probably correct, although he could not suggest a possible ancestor among the birds of the South American continent. The characters which showed the clearest interspecific differences were beak dimensions (as noted 100 years earlier by Darwin). Lack considered these differences as adaptations to different kinds of food, mainly seeds. When two species inhabited the same island, the distributions of their beak characters were wider apart than when each species inhabited a different island. This indicated possible competition for food as a reasonable driving force for evolution by natural selection.

Two American scientists, P.R. and B.R. Grant, studied the finch populations on two Galapagos islands for more than 20 years. They banded many birds individually, measured and weighed them and followed their fate and that of their offspring. There were two climatically exceptional years during the research period ('El Ninio' years) – 1976/7 and 1983/4 – with drastic changes in weather and vegetation. Only 14% of the birds survived the first event, and 32% survived the second (Grant and Grant, 1995 a, b).

The distribution of beak size in the offspring was predictable from the sizes of available seeds after each weather change (Boag and Grant, 1981). After the first El Ninio, the survivors were mainly the largest birds with the largest beaks. This was reflected in the distribution of beak sizes in their offspring (Grant, 1991). Among the survivors of the second event, those with the smallest beaks were particularly numerous (Grant and Grant, 1993). The second El Ninio caused great damage to the cacti (*Opuntia*), which provide the main food source for the birds with the largest beaks, but the heavy rains had been favorable to grasses and there was an abundance of small grass seeds (Grant and Grant, 1989).

Literature Cited

Berry, R.J. 1990. Industrial melanism and peppered moths (*Biston betularia* (L.)). Biological Journal of the Linnean Society 39: 301-322.

Bishop, J.A. 1972. An experimental study of the cline in industrial melanism in *Biston betularia* (L.) (Lepidoptera) between urban Livepool and rural North Wales. Journal of Animal Ecology 41: 209-243.

Boag, P.T. and P.R. Grant. 1981. Intense natural selection in a population of Darwin's Finches (Geospizinae) in the Galapagos islands. Science 214: 82-85.

Bradshaw, A.D. 1971. Plant evolution in extreme environments. pp. 20-50. In: R. Creed, (ed.), Ecological Genetics and Evolution. Blackwell, Oxford, UK.

Brakefield, P.M. 1987. Industrial melanism: do we have all the answers? Trends in Ecology and Evolution 2: 117-122.

Brown, A.W.A. 1960. Mechanisms of resistance against insecticides. Annual Review of Entomology 5: 301-326.

Cameron, R.A.D. and D.J. Pannett. 1985. Interaction between area effects and variation within habitats in *Cepaea*. Biological Journal of the Linnean Society 24: 365-379.

Cameron, R.A.D. 1992. Change and stability in *Cepaea* population over 25 years: a case of climatic selection. Proceedings of the Royal Society, London B 248: 181-187.

Clarke, C.A., G.S. Mani and G. Wynne. 1985. Evolution in reverse: clean air and the peppered moth. Biological Journal of the Linnean Society 26: 189-199.

Cook, L.M., G.S. Mani and M.E. Varley. 1986. Post-industrial melanism in the peppered moth. Science 231: 611-613.

Cook, L.M., K.D. Rigby and M.R.D. Seaward. 1990. Melanic moths and changes in epiphytic vegetation in north-west England and north Wales. Biological Journal of the Linnean Society 39: 343-354.

Creed, R. 1971. (ed.) Ecological Genetics and Evolution. Blackwell, Oxford, UK.

Devonshire, A.L. 1989. Insecticide resistance in *Myzus persicae*: from field to gene and back again. Pesticide Science 26: 375-382.

Devonshire, A.L. and L.M. Field. 1991. Gene amplification and insecticide resistance. Annual Review of Entomology 36: 1-23.

Endler, J.A. 1984. Progressive background in moths and a quantitative measure of crypsis. Biological Journal of the Linnean Society 22: 187-231.

Field, L.M., A.L. Devonshire and B.G. Forde. 1988. Molecular evidence that insecticide resistance in peach-potato aphids (*Myzus persicae* Sulz.) results from amplification of an esterase gene. Biochemical Journal 251: 309-312.

Firko, M.J. and J.L. Hayes. 1990. Quantitative genetic tools for insecticide resistance risk assessment. Journal of Economic Entomology 83: 647-654.

Ford, E.B. 1964. Ecological Genetics. Methuen, London, UK.

Forgash, A.J. 1984. History, evolution, and consequences of insecticide resistance. Pesticide Biochemistry and Physiology 22: 178-186.

Goodhart, C.B. 1987. Why are some snails visibly polymorphic, and others not? Biological Journal of the Linnean Society 31: 35-58.

Grant, P.R. 1991. Natural selection and Darwin's Finches. Scientific American, October 1991, pp. 660-665.

Grant, B.R. and P.R. Grant. 1989. Natural selection in a population of Darwin's finches. American Naturalist 133: 377-393.

Grant, B.R. and P.R. Grant. 1993. Evolution of Darwin's Finches caused by a rare climatic event. Proceedings of the Royal Society, London B 251: 111-117.

Grant, P.R. and B.R. Grant. 1994. Phenotypic and genetic effects of hybridization in Darwin's finches. Evolution 48: 297-316.

Grant, P.R. and B.R. Grant. 1995a. Predicting micro-evolutionary responses to directional selection on heritable variation. Evolution 49: 241-251.

Grant, P.R. and B.R. Grant. 1995b. Founding of a new population of Darwin's finches. Evolution 49: 229-240.

Grant, B.S., D.F. Owen and C.A. Clarke. 1996. Parallel rise and fall of melanic peppered moths in America and Britain. Journal of Heredity 87: 351-357.

Hagen, J.B. 1999. Retelling experiments: Kettlewell's studies of industrial melanism in peppered moths. Biology and Philosophy 14: 39-54

Huxley, J. 1942. Evolution. The Modern Synthesis. George Allen & Unwin, London, UK.

Johnson, M.E., O.C. Stine and J. Murray. 1984. Reproductive compatibility despite large-scale genetic divergence in *Cepaea nemoralis*. Heredity 53: 655-665.

Jones, J.S., B.H. Leith and P. Rawlings. 1977. Polymorphism in *Cepaea*. Annual Review of Ecology and Systematics 8: 109-143.

Lack, D. 1947. Darwin's Finches. Cambridge University Press, Cambridge, UK.

Kettlewell, H.B.D. 1973. The Evolution of Melanism in the Lepidoptera. Oxford, UK.

Kettlewell, H.B.D. and R.J. Berry. 1961. The study of a cline: *Amathes glareosa* Esp. and its melanic f. edda Stand (Lep.) in Shetland. Heredity 16: 404-413.

Lees, D.R. and E.R. Creed. 1975. Industrial melanism in *Biston betularia*: the role of selective predation. Journal of Animal Ecology 44: 67-83.

Lees, D.R., E.R. Creed and J.G. Duckett. 1973. Atmospheric pollution and industrial melanism. Heredity 30: 227-232.

Liebert, T.G. and P.M. Brakefield. 1987. Behavioral studies on the peppered moth *Biston betularia* and a discussion of the role of pollution and lichens in industrial melanism. Biological Journal of the Linnean Society, 31: 129-231.

Macnair, M.R. 1991. Why the evolution of resistance to anthropogenic toxins normally involves major gene changes: the limits to natural selection. Genetica 84: 213-219.

Mani, G.S. 1980. A theoretical study of morph ratio clines with special reference to melanism in moths. Proceedings of the Royal Society, London B 210: 299-316.

McKenzie, J.A. 1984. Dieldrin and diazinon resistance in populations of the Australian sheep blowfly, *Lucilia cuprina*, from sheep grazing and rubbish tips. Australian Journal of Biological sciences 37: 367-374.

McKenzie, J.A., A.G. Parker and J.L. Yen. 1992. Polygenic and single gene responses to selection for resistance to diazinon in *Lucilia cuprina*. Genetics 130: 613-620.

Mikkola, K. 1979. Resting site selection of *Oligia* and *Biston* moths (Lepidoptera: Noctuidae and Geometridae). Annales Entomologici Fennici 45: 81-87.

Mikkola, K. 1984. On the selective forces acting in the industrial melanism of *Biston* and *Oligia* moths (Lepidoptera: Geometridae and Noctuidae). Biological Journal of the Linnean Society 21: 409-421.

Mouches, C., N. Pasteur, J.B. Berge, O. Hyrien, M. Raymond, R.B. Saint Vincent, M. De Silvestri and G.P. Georghiou. 1986. Amplification of an esterase gene is responsible for insecticide resistance in a California *Culex* mosquito. Science 233: 778-780.

Pasteur, N. and M. Raymond. 1996. Insecticide resistance genes in mosquitoes: their mutation, migration and selection in field populations. Journal of Heredity 87: 444-449.

Perry, A.S. and M. Agosin. 1974. The physiology of insecticide resistance by insects. pp. 3-121. In: M. Rockstein (ed.) The Physiology of Insecta, vol. VI.

Pimentel, D. and A.C. Bellotti. 1976. Parasite-host systems and genetic stability. American Naturalist 110: 877-888.

Pinniger, D.G. 1974. The behavior of insects in the presence of insecticides: the effect of fenitrothion and malathion on resistant and susceptible strains of *Tribolium castaneum*. In: Proceedings, Working Conference on Stored-Products Entomology, Savannah, Ga. USA, pp. 301-308.

Raymond, M., A. Callaghan, P. Fort and N. Pasteur. 1991. Worldwide migration of amplified insecticide resistance genes in mosquitoes. Nature 350: 151-153.

Reznick, D.N. and C. Ghalambor. 2001. The population ecology of contemporary adaptation: what empirical studies reveal about the conditions that promote adaptive evolution. Genetica 112-113: 183-198.

Reznick, D.N. and C. Ghalambor. 2005a. Can commercial fishing cause evolution? answers from guppies. Canadian Journal of Fisheries and Aquatic Science 62: 791-801.

Reznick, D.N. and C. Ghalambor. 2005b. Selection in nature: experimental manipulation of natural populations. Integrative and Comparative Biology 45: 456-462.

Reznick, D.N., M.J. Butler and H. Rodd. 2001. Life history evolution in guppies. VII. The comparative ecology of high and low predation environments. American Naturalist 157: 126-140.

Roush, R.T. and J.A. McKenzie. 1987. Ecological genetics of insecticide and acaricide resistance. Annual Review of Entomology 32: 361-380.

Roush, R.T. and J.C. Daly. 1990. The role of population genetics in resistance research and management. pp. 97-152. In: R.T. Roush and B.E. Tabashnik (eds.), Pesticide Resistance in Arthropods. Chapman & Hall. New York, USA.

Schweiger, O., M. Frenzel and W. Durka. 2004. Spatial genetic structure in a meta-population of the land snail, *Cepaea nemoralis* (Gastropoda: Helicidae). Molecular Ecology 13: 3645-3655.

Silverman, J. and M.H. Ross. 1994. Behavioral resistance of field-collected German cockroaches (Blattodea: Blattellidae) to baits containing glucose. Environmental Entomology 23: 425-430.

Sokal, R.R. and F.J. Rohlf. 1995. Biometry. 3rd ed. Freeman, New York, USA.

Vekemans, X. and C. Lefebvre. 1997. On the evolution of heavy metal tolerant populations in *Armeria maritima*: evidence from allozyme variation and reproductive barriers. Journal of Evolutionary Biology 10: 175-191.

West, D.A. 1977. Melanism in *Biston* (Lepidoptera: Geometridae) in the rural central Appalachians. Heredity. 39: 75-81.

Wool, D. and O. Manheim. 1980. Genetically-induced susceptibility to malathion in *Tribolium castaneum* despite selection for resistance. Entomologia Experimentalis et Applicata 28: 183-191.

Wool, D. and S. Noiman. 1983. Integrated control of insecticide resistance by combined genetic and chemical treatments: a warehouse model with flour beetles (*Tribolium*. Tenebrionidae: Coleoptera). Zeitschrift der Angewandte Entomologie 95: 22-30.

Wool, D., S. Noiman, O. Manheim and E. Cohen. 1982. Malathion resistance of *Tribolium* strains and their hybrids. Biochemical Genetics 20: 621-636.

Wool, D., R. Aloni, O. Ben Zvi and M. Wollberg. 1999. A galling aphid furnishes its home with a built-in pipeline to the host food supply. Entomologia Experimentalis et Applicata 91: 183-186.

Wu, L., A.D. Bradshaw and D.A.Thurmann. 1975. The potential for evolution of heavy metal tolerance in plants. III. The rapid evolution of copper tolerance in *Agrostis tenuis*. Heredity 34: 165-187.

CHAPTER **13**

Natural Selection and Adaptation

WHAT IS ADAPTATION?

Organisms do not live in a vacuum. They absorb and use materials from the environment (oxygen, food, water and minerals), excrete various compounds into it (CO_2, digested materials, faeces and urine) and leave their own decaying tissues in it when they die. Some organisms physically change the habitat in very conspicuous ways: termites build huge mounds for their nests; Earthworms (*Lumbricus* spp.) transport quantities of earth from the depth of their burrows to the surface (Darwin 1881); mole rats, ground squirrels and ants dig underground networks. An extreme in magnitude – but not unique in principle – is the effect of man on the global environment.

The environment continuously presents the organisms with challenges to which they must respond: drought, floods, temperature changes, predators, parasites and competitors. Organisms exist and flourish in habitats which appear (to humans) as extremely harsh – like hot and dry deserts or the extreme cold of the Arctic and Antarctic. It is reasonable to expect that these organisms have evolved characters which enable them to survive and reproduce, whereas other organisms are unable to exist in such environments. These characters are referred to as adaptive: the organisms are adapted to these habitats.

[Some authors advocate that the term 'adaptive' should be restricted only to characters which can be assumed to have evolved via natural selection (e.g., Gottard and Nylin, 1995). Where this cannot be assumed, they suggest that the trait should be referred to as 'beneficial' (in contrast with 'non-beneficial' or 'maladaptive'.]

Adaptation is sometimes expressed in morphological characters. Some succulent desert plants, like cacti, store water in their tissues in the brief rainy season. Others lose their leaves or replace them with thorns, reducing transpiration and water loss. Some of the best examples of adaptation in plants are found in the complex structures of the flowers in orchids, which serve to direct the pollinators to the nectar at the right position for the transfer of the pollinia to another flower (Darwin, 1984 (1877)). Detailed investigation of the morphology, physiology, or behavior of any organism will reveal evidence of adaptation.

The skeletal bones in birds are hollow. This is considered an adaptation for flight – by reducing skeleton weight without sacrificing bone strength. In fishes, the gills, swim bladder, hydrodynamic body shape and tail are adaptations for life in the aquatic medium. The legs of the praying mantis are adaptations for capturing prey, those of the mole cricket - for digging in the soil. Nevo (1979) described in detail the adaptations of the mole rat (*Spalax*), a subterranean mammal, for life underground – including its head shape, rudimentary eyes, short legs, bi-directional fur and physiology.

Physiological research in desert animals like the camel (Schmidt-Nielsen, 1964) and the black Beduin goat (Shkolnik et al., 1979) revealed complex adaptations to the shortage of water, notably the ability to replenish quickly the water deficit after a prolonged period without drinking. Adaptations in beetles of the Namib desert include drinking the droplets of dew which form on cold surfaces early in the morning (including on their own cuticle). Other desert organisms 'escape' from unsuitable conditions by nocturnal activity or dormancy. Many more examples are listed by Huxley (1942, p. 420 ff).

Parasites and hosts

Many examples of close adaptation of parasites to their hosts are provided by gall-forming insects. The host-specific aphids feeding on *Pistacia* (Fordinae: Aphidoidea, Pemphigidae) illustrate the phenomenon (Wool, 2004; Inbar and Wool, 1995). Only the fundatrix nymph, emerging from an overwintering egg, is able to induce a gall. Gall shape is insect species-specific, and is induced on a specific host tree (gall photographs are shown in Wool, 2004). The fundatrix reproduces parthenogenetically in the gall, and hundreds or even thousands of genetically-identical daughters are produced (the daughters are unable to induce galls of their own!). The fundatrix induces dramatic changes in the structure of the plant vascular system – which enable the numerous phloem-feeding daughter aphids to use the resources produced by photosynthesis in the plant (a case of 'environmental engineering'; Wool et al., 1999).

The concept of <u>pre-adaptation</u> (which means that the population is genetically prepared to face deleterious conditions before they actually occur), is controversial. Goldschmidt (1933) explained the appearance of 'hopeful monsters' by slow accumulation of small, mostly harmless mutations, which may produce a phenotype pre-adapted to new conditions – and perhaps be the start of a new species (but will be selected against as a 'monster' if the environment remains unchanged).

Do We Need the Term 'Adaptation'?

Krimbas (1984) argued that all the definitions of adaptation are phrased in terms of fitness, and they are in fact cases of circular reasoning: "selective processes bring about adaptation because the survivors – the better fit – are the most adapted, and they have the greatest probability of surviving and reproducing". The term adaptation is a remnant from pre-Darwinian times when each phenomenon and organ was considered to have a 'purpose' or a predetermined destiny, and each organ as best planned to fulfil its function. To identify a character as adaptive, Krimbas argues, we must check how well it fulfils its function. This is a teleological approach: the function is something that the observer can decide upon only from observing the character. It is clear that the function of the eye is to see, and from that we understand the functions of the parts of the eye structure, like the retina and the lens. This explanation is no different from Paley's explanation in 'Natural Theology' (Chapter 1) except that the role of the Creator is taken over by natural selection. Krimbas suggests that the term adaptation be dropped, and the results of selection stated in fitness terms alone.

For arguments to the contrary, see Wallace (1984).

THE EVOLUTION OF ADAPTATION

Slobodkin (1968) described a model for the evolution of adaptation. The model assumes that every organism is at equilibrium with the surrounding environment: a change in the environment disturbs the balance and is potentially deleterious. In every individual there exists a hierarchy of possible responses to change, attempting to restore the balance and prevent damage, with a minimal loss of energy.

In animals, the first option is a behavioral response. Imagine a dog lying in an exposed spot in front of its kennel in the morning. As the sun rises, so does the ambient temperature. To function properly, the mammalian body temperature must remain within limits: if exceeded, it may damage the dog's internal environment. Before this point is reached, the dog will get up and move to a shaded site. This response is the 'cheapest' energetically.

If the balance is not restored and temperature still increases, the level next in the hierarchy – the physiological level – is used. The dog will open its mouth, stick out its tongue and pant. Evaporation from the lungs and buccal cavity absorbs the heat from the body and cools it down to restore the balance. But panting requires intensive muscular activity, and has the disadvantage of losing water. Physiological mechanisms are more costly energetically than behavioral ones, and there must be a limit to the time this activity can continue without draining the dog's resources.

Physiological adaptations for maintaining temperature balance are known. Fledglings of swifts in cold countries can lower their metabolism, and body temperature drops almost down to ambient temperature when their parents leave them in search for food and are unable to keep them warm (birds have a body temperature in excess of 40°C). Lowering the metabolic rate conserves energy.

The natives living at high altitudes on the Andes are adapted to the lower oxygen levels and have more red cells in their blood. Athletes who arrived in Mexico City for the Olympic games found it difficult to maintain their achievements (Mexico City is situated at 2000 m elevation): their hearts had to work harder to supply enough oxygen to the straining muscles. After a few weeks' stay at the high altitude, more red-blood corpuscles were produced in their blood and their performance improved.

If the physiological mechanisms fail to restore the balance with the environment, the individual will die. This is the point where natural selection begins to work, and adaptation may result.

Physiological characteristics of organisms must be determined genetically to some extent. Imagine a genetically variable population, in which some individuals have a wider range of heat tolerance than others. Alleles affecting metabolic rates, nerve function, and other physiological traits may likewise be variable among individuals. If a long-term environmental change takes place, some organisms will die, but mortality will not be random. Just as in the case of insecticide resistance, those genetically endowed with the greater tolerance may have a better chance of surviving and will pass on their genes to their offspring. If the process is repeated, natural selection will adapt the population to the new conditions.

Example

Evolution of an adaptive behavioral response in Drosophila

A behavioral change may allow a population to survive in a deleterious environment, or to use a previously unexploited resource.

Four large populations of *Drosophila willistoni* – numbering thousands of flies – were established in order to investigate the effects of radiation (De Souza et al., 1970). The population cages contained periodically-replaced food cups, in which the larvae developed to adulthood. To prevent drying of the food, the cages were held in a humid environment. Unrelated to the original purpose of the experiment, the investigators noticed a change in the behavior of the pupating larvae. Normally the larvae pupated on the walls of the food cups, near the food surface, and only a few crawled out and pupated on the floor cage – and these usually desiccated and perished. But as the experiment continued in the humid environment – about 130 generations – more and more pupae were formed on the floor of the cage, and many of them hatched as adults (Fig. 13.1).

A short-term selection experiment revealed that the tendency to pupate 'in' or 'out' of the food was heritable: There was very fast response to selection in either direction (Fig. 13.1, bottom).

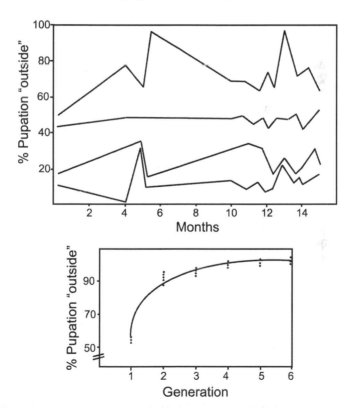

Fig. 13.1 Evolution of a behavioral response to environmental change in laboratory populations of *Drosophila willistoni* (after De Souza et al., 1970). Top: percentage pupation 'outside' in the four populations. Bottom: artificial selection for pupating 'out' of the medium.

Genetic crosses showed that the tendency to pupate 'outside' was due to a single, dominant allele. The carriers of this trait remained near the food when the humidity in the cage was low, but in the experimental humid environment, more of them survived when they crawled out of the crowded food cups. In doing so they escaped competition for the limited pupation sites on the walls of the food cups.

MEASURING ADAPTATION

Julian Huxley (1942) was one of the advocates of natural selection as the mechanism bringing about adaptation. He based this conclusion on the 'instincts' of the field naturalist, and scorned those who did not believe in the selective and adaptive value of protective coloration, mimicry or threat behavior (Huxley, 1942, p. 413). 'Naturalist instinct' alone, however,is not sufficient to support the claim that adaptation is brought about by selection: quantitative data are required.

To measure the process of adaptation, genetic markers are needed which are associated with fitness-related genes, and transmitted from one generation to the next as a unit, without recombination, so that the frequencies of the markers enable tracing of the changes in fitness. Such markers are hard to find. It is not easy to show that selection is making the population better adapted, or that one population is better adapted than another, because there are no objective, quantitative ways to measure adaptation.

Fitness Components

In theoretical studies, it is convenient to use W – the 'Darwinian fitness'. This parameter increases with the removal of deleterious alleles from the population, as expected from the process of adaptation (Chapter 9). But W does not measure adaptation to the environment: it is a relative measure of reproductive output of genotypes, relative to the most prolific one, regardless of the environment.

Intuitively, it is reasonable to assume that a better-adapted population would have a higher fitness in the particular environment than a less-well-adapted population: allowing enough time, better adapted genotypes will be selected, so that the average fitness in the population should increase. It was suggested that fitness components like survival, reproductive output, larval or pupal weights (in insects), and the length of the developmental period could be used as measures of adaptation.

Dawson and Riddle (1983) measured fitness components in experimental populations of the flour beetle, *Tribolium castaneum,* when transferred to new environments. Environments were defined as new kinds of food medium (flour beetles mate, oviposit, and spend their entire life as immatures and adults in the medium, except for brief periods of dispersal). After 60 generations in the new media, the investigators found no indication of adaptation as measured in fitness parameters. On the contrary, the selected populations appeared less fit in the new media than samples from the unselected source population.

The intrinsic rate of increase

The ecological intrinsic rate of population increase *r* was suggested as a measure of adaptation. In a constant environment with fixed resources, the fastest growing population has the greatest chance of dominating the habitat (equating growth rate with competitive ability). Different populations in the same environment can then be compared.

The intrinsic rate of increase combines two population parameters: fecundity per female and generation time (high fecundity and short generation time increase *r*). Lewontin (1965) showed that generation time is the more important of the two. In fact, it is often easier to select for higher fecundity – in an experimental population – than to reduce developmental time.

However, *r* is not always a good measure of adaptation. A higher growth rate does not always imply better fitness: If the environment – particularly the food supply – is limited, or is not renewed fast enough, a high *r* may cause early extinction (e.g., Wool and Sverdlov, 1976).

Ecologists found that small body size, short developmental time and high fecundity – characteristics of *r*-strategists – are useful in recovery after a catastrophe, where the balance with the environment had been disrupted. Species with high *r* often occur in unstable, ephemeral and unpredictable habitats. Success in such environments is not indicative of fitness in the long run.

Resource utilization

The ability to convert environmental resources into population biomass may be a measure of adaptation. A population is better adapted if it can make use of a larger proportion of the available resources (Ayala, 1969).

Adaptation ('adaptedness' in Ayala's terminology) of different populations may be compared in the laboratory by measuring the amount of biomass (total weight or population size) produced from a fixed amount of food in a fixed time interval, after the populations reached a stable equilibrium level, where the numbers of newborn added equal the numbers of dead removed. Ayala (1969) suggested an experimental system for measuring the relative adaptedness of different *Drosophila* populations: the adults are transferred periodically to new media where their eggs develop adults (the 'serial transfer' method). This method was tried with flour beetles (Wool, 1973) but no equilibrium was reached: the serial-, transfer method seems to be organism-dependent.

Adaptation to Adverse Conditions?

Situations where conditions become temporarily worse – gradually or suddenly – are probably common in nature. Climatic changes, infection by fungi or bacteria, the introduction of competitors, or the extensive use of resources by the growing population may reduce the amount of food or space, or make it unusable through dehydration or excretion of waste products. Individuals which can use less-than-optimal food (or smaller quantities) and still reach adulthood and reproduce may have an advantage over other individuals. Can natural selection bring about adaptation in the sense of ability to survive adverse conditions?

In *Drosophila* (Robertson, 1965) and in houseflies (Sullivan and Sokal, 1963) development is normally prolonged until the larva accumulates a critical amount of biomass. The ability to pupate at a lower-than-normal weight may be advantageous in situations of limited resources.

When housefly larvae were removed from food after reaching the critical weight, they pupated without further feeding. When removed from food before reaching the critical weight, they died. Body weight in insects is directly related to fecundity: larger females lay more eggs. At high densities (less food per larva) the proportion of larvae reaching the adult stage may not be affected, but the resulting adults will be smaller than normal. On the other hand, these smaller adults may disperse and find new and better oviposition sites, where their (fewer) eggs will have a chance to develop with less competition.

CO-ADAPTATION AND CO-EVOLUTION

There are many examples of close adjustment and interdependence of the life cycles between animals and plants, which also involve particular morphological, anatomical and physiological mechanisms. Adjustments between parasites and hosts, predators and prey, and between competing species, have been described in detail in the ecological literature. Such cases are described as **co-adaptation**. The process that brings them about is called **co-evolution**.

For example, the life cycles of the tiny wasps of the family Agaonidae are closely adjusted with the reproductive cycle of fig trees (*Ficus* spp.). The wasps develop in the inflorescences (figs) and pollinate the flowers (Wiebes, 1979; Galil and Eisikowitch, 1968, 1970). Minute details of the wasp anatomy and behavior facilitate its entering the fig at the right time, ovipositing in some of the flowers (which develop into galls harboring the wasp larvae) and pollinating the rest.

The term co-evolution is used in the literature for rather different situations. One common usage is to describe cases of apparently parallel phylogenetic trees of the host plants and the insects (e.g., butterflies). Examples are described in the book edited by Futuyma and Slatkin (1983). Taxonomically closely-related species of butterflies (e.g., of the family Papilionidae; Scriber, 1983) frequently feed on closely-related plants containing the same alkaloids ('secondary compounds') that may be toxic to unrelated insects.

The mechanism bringing about co-evolution was described as a kind of arms race (Ehrlich and Raven, 1964). A mutation in the insect population creates a new genotype, which is able to cope with a plant alkaloid. The presence of this alkaloid prevents the use of the plant by non-mutated individual insects. The new genotype has a great advantage since it can use a host plant with no competitors. The damage to the plant increases until a mutation in the plant population results in a changed alkaloid, which makes mutant plant genotypes immune to the insect. The insect advantage is reduced, until another mutation occurs, and so the cycles continue.

An arms-race scenario is a likely description of the case of the Hessian fly *Mayetiola destructor* (Cecidomyiidae), a midge pest of wheat. The female lays its eggs in wheat stems. As a result of larval activity, the stems bend and fall before harvest, causing heavy damage to farmers. Wheat-breeding research has yielded several varieties resistant to the midge (Gallun, 1977; Foster et al., 1991), but each time a fly 'biotype' emerged which overcame the plant genetic resistance barrier. This is in fact an arms race between the Hessian fly and wheat (mediated by man), in which each combatant evolves new genotypes to deal with the changes in the other.

Other cases of co-evolution may also have evolved as an arms race. One such case is the co-adaptation of the Yucca plant which grows in the deserts of central America and southern USA, with specific moths (*Tegeticula*, family Prodoxidae). The Yucca plant produces a tall column with several hundred large, white flowers. The antennae of the pollinating moth are specially adapted for placing pollen on the stigma of the plant, and the female moth pollinates the flower before ovipositing into the ovary. Like the fig wasps and figs, the relationship is obligatory for both the plant and the insect: the plant drops its flowers and fruit-pods unless fertilized, and the insect larvae feed on the seeds in the pod – but they must not eat too many seeds too early to avoid starvation. Some seeds remain uneaten for plant propagation (Aker and Udovic, 1981; Thompson and Pellmyr, 1992; Ziv and Bronstein, 1996).

Can Competition Bring about Co-adaptation?

If competitive interactions among species in nature sometimes end in co-adaptation rather than competitive exclusion, it would explain the evolution of plant and animal communities. In the multi-dimensional ecological niche model – where each environmental factor constitutes a dimension – each species' niche is delimited by its requirements of each dimension. A species may perhaps share some of these requirements with other species. Overlap of requirements creates competition, when the dimension represents a limiting resource (e.g., space or food). If each species limits its requirements as a result of competition, the overlap may be reduced to the benefit of both contestants. In evolutionary terms, this constitutes co-adaptation. The limiting of requirements means a change in the genetic composition of one or both populations. Can natural selection bring about such an adjustment?

An experiment simulating such a natural process was carried out by Seaton and Antonovics (1967). They reared two *Drosophila* strains together in the larval stage, but prevented their mating with each other, so that no inter-strain hybrids were formed. Competitive ability of the two strains was tested every generation in samples extracted from the mixed populations, separated by strain and reared to adults. The results showed that after five generations, one of the two strains had improved its ability to exist in the mixed culture – while the second strain showed no response. It seems however, that this result was atypical. Negative results were reported in experiments with 25 of 28 other combinations of *Drosophila* strains (Futuyma, 1970). Similar experiments with houseflies and flour beetles showed no response either (Sokal et al., 1970). Although it is tempting to accept that natural selection can bring about co-adaptation, the limited data available so far provide no support for the idea.

NATURAL SELECTION AND GEOGRAPHICAL CLINES

When examined over a large geographic area, or even on a global scale, morphological, physiological or other characters in natural populations are often correlated with variation in some environmental parameter, such as latitude, elevation, temperature, rainfall or day-length. The term 'cline' for such geographical correlations was suggested by Julian Huxley (Mayr, 1942, p. 95). Indeed, Huxley (1942) lists many examples of clines. Three of the more famous ones are referred to as 'rules'. **'Bergmann's rule'** states that in homeothermic animals, body size (within species) increases as the habitat temperature decreases (e.g., when one travels northwards in North America). **'Allen's rule'** states that the relative lengths of legs, tail and ears become shorter as the temperature decreases (these limbs are the longest in hot-desert animals). **'Gloger's rule'** states that the amount of dark pigmentation (e.g., melanin) in the skin decreases with temperature and increases with humidity (the darkest specimens occur in the hot and humid tropics) (Huxley, 1942, p. 213).

A recent example of a cline comes from a study of an Australian marsupial, the common brush-tailed possum, which was introduced to New Zealand in 1837 and became widespread in the two large islands. Measurements of 300 skulls collected along the possum's distribution area showed a clear negative correlation with latitude – northern animals had larger skulls – as predicted from Bergmann's rule (Yom Tov et al., 1986). This cline must have developed in less than 150 years!

Huxley and Mayr explain that such clines were formed by natural selection. The increase in body size and the shortening of legs and ears as mean temperatures drop, should be advantageous for physiological reasons (in homoeothermic animals), because the larger the animal – the smaller is the ratio of surface area to volume – thus reducing heat loss, and helping the animal maintain its constant body temperature. A small individual must spend more energy in keeping warm (the smallest mammals, the shrews, must consume almost their body weight in food daily to survive even in temperate climates).

This explanation – which is intuitively acceptable to every biologist and often cited – has nevertheless never been properly proved (to the best of my knowledge). The observed clines could result from factors unrelated to the measurable environmental variable such as temperature. Body size is a quantitative character, directly affected by environmental as well as genetic (physiological, hormonal) factors, which can be independent of natural selection. These environmental effects are difficult to control even in laboratory experiments.

> The following information was published in 1924, on white laboratory rats propagated by H. Przibram in Vienna:
> "... Many of the exterior parts of the body show a stimulated growth, even though the body of rats kept at a high temperature, compared to rats kept at a low temperature, as a whole, is noticeably reduced in size. The rats kept at a high temperature have a more pointed mouth, bigger ears, and longer, narrower paws. They have a decidedly enlarged scrotum and, finally, very noticeably long tails" (Kammerer 1924, p. 324-325).

These doubts are supported by data on clines, consistent with Bergmann's rule, in poekilothermic animals which do not need to maintain a constant body temperature. Ray (1960) lists 17 cases of reptiles and insects where such clines were reported.

Wool (1977) studied samples of gall-inducing aphids, from 35 sites in Israel and the Sinai. A negative correlation was found between size-related morphological characters of the aphid *Geoica wertheimae* (referred to as *G. utricularia*) and the mean annual temperature at the collection site. It is unclear what selective advantage larger aphids might have at low temperatures.

A selective advantage for larger insects in northern latitudes could be explained by factors other than temperature adaptation. Masaki (1978) studied a cricket which is distributed across a very wide geographical range – from 24° to 44° of latitude – in the islands of Japan. Mean temperatures vary from 5°C in the north to 24°C in the south of its distribution. Head width was used as a criterion of cricket size. Between 24° and 32° latitude, there was no substantial change in cricket size – while above and below this geographical segment the size conformed to Bergmann's rule. Masaki explains that the cline in the northern and the southern parts of the range resulted from the effect of temperature and reduced daylight duration on developmental time and the timing of diapause. In the middle part of the range, more than one generation was completed before the onset of winter. These factors, and their indirect effects on fecundity, shaped the cline rather than direct temperature dependence.

Geographical Clines Resulting from Historical Processes

A wide-scale electrophoretic study of the gall wasp, *Andricus quercuscalicis*, on oaks in Europe revealed a cline – which was most probably created by a historical process and not from physiological adaptation (Stone and Sunnucks, 1993).

This gall wasp requires two host trees to complete its annual life cycle. An asexual generation develops on the acorns of the pedunculate oak, *Quercus rubor*, and a sexual generation on the flowers of Turkey oak, *Q. cerris*. Turkey oak disappeared from Europe during the ice ages but survived in southern refugia in Turkey and in Spain. It was introduced and replanted in the last 300-400 years in parks and commercial forests throughout Europe, for timber and shade, and the dates of these introductions are on record. The wasp must have followed these introductions, which enabled it to complete the life cycle and attack the acorns on *Q. rubor* (Fig. 13.2).

Stone and Sunnucks (1993) surveyed 39 populations of the wasp across Europe, from Turkey to Britain. They found significant differentiation – and a linear loss of genetic variation over distance. Sites in the native range of the Turkey oak, where the wasp must have existed for the longest period of time, were the most variable. In samples from western Europe,

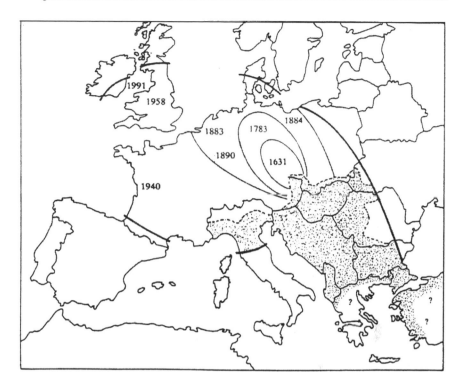

Fig. 13.2 Tracing of the historical records of the distribution of Turkey Oak by reforestation and planting in parks. The gall wasp, *Andricus quercuscalicis*, followed the reintroduction of its secondary host tree which enabled it to complete its life cycle (from Stone and Sunnucks, 1993; reproduced with permission of the publisher).

which were colonized most recently, polymorphism was low, tending to monomorphism.

Stone and Sunnucks regard this cline as "the most dramatic and linear loss of genetic variation over distance yet demonstrated in an invading animal". After a careful examination of alternative explanations, Stone and Sunnucks conclude that the loss of genetic variation in this clinal manner was probably the result of genetic drift and founder effects during the directional migration of the wasps westwards from the refugium in the east, as the host trees became available – not by natural selection.

NON-SELECTIVE EXPLANATIONS FOR ADAPTATION

How do we identify an adaptive trait? Decisions as to whether a trait is adaptive are based on comparisons and associations: for example, giraffes have long necks, buffalos have short necks. Buffalos feed on grass, giraffes on tree leaves. Conclusion: the long neck of the giraffe is an adaptation for feeding on trees.

These are post-hoc explanations. There is no information on how either of these organisms evolved these traits. There is no way of assigning cause and effect. The differences between the two species could result from growth and development patterns independent of feeding habits.

Some believe that natural selection has optimized everything during the millions of years of evolution. Each and every character is supposed to be in its optimal state. The organism's fitness is a kind of a sum of the fitness conferred by all its characters. But the state or use of a character today cannot tell us how or why it evolved to be that way: there are different, perhaps equally efficient ways of reaching the present state of adaptation to a given environment (and past environments may not have been like the present one). The evolution of the horse's foot, from four toes to a single one, is considered an adaptation for faster running – but the cheetah, with five toes on its forepaws and four on the hind paws, can run faster than the horse (Fig. 13.3).

The 'adaptionist paradigm', which assumes that the existing character state is a selective optimum for each trait, was severely criticized by Gould and Lewontin (1979). Characters can exist in their present state because they were inherited that way from an ancestor, or due to ontogenetic constraints in embryology, or due to a chance mutation that was fixed in a population in the remote past (all mammals have four legs not because this is a selected optimum, but because mammals evolved from reptiles that possessed four legs). The fact that the trait <u>could</u> have evolved by natural selection is no proof that it did. As Darwin himself wrote in the 'conclusions' of his book:

> I am convinced that natural selection has been the main, but not the exclusive, means of modification (Darwin, The Origin of Species, Conclusions)

Adaptation and 'the Red Queen hypothesis'

The Red Queen in Lewis Carroll's 'Through the Looking Glass' explains that you must keep running to stay where you are. It was used by Van Valen (1973) as a metaphor, to mean that adaptation is a continuous process. Populations – or species – must be modified continuously by natural selection, in order to keep pace with the changing environment. Species which cannot keep up with the changing environment will become extinct.

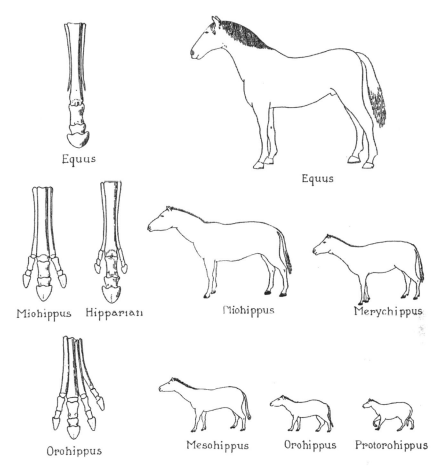

Fig. 13.3 Changes in body size and the anatomy of the foot in the evolution of the horse (from Morgan, 1925; reproduced with permission of Princeton University Press).

Literature Cited

Aker, C.L. and D. Udovic. 1981. Oviposition and pollination behavior of the yucca moth and its relation to the reproductive biology of yucca. Oecologia 49: 96-101.

Ayala, F.J. 1969. An evolutionary dilemma: fitness of genotypes versus fitness of populations. Canadian Journal of Genetics and Cytology 11: 439-456.

Darwin, C. (1881). 1985. The formation of vegetable mould through the action of worms. University of Chicago Press, USA.

Darwin, C. 1889. The Origin of Species, 6th ed. Murray, London, UK.

Darwin, C. (1984) (1877). The Various Contrivances by which Orchids are Fertilized. 2nd ed. University of Chicago Press, Chicago, USA.

Dawson, P.S. and R.A. Riddle. 1983. Genetic variation, environmental heterogeneity, and evolutionary stability. pp. 147-170. In: C.E. King and P.S. Dawson (eds.) Population Biology: Retrospect and Prospect, Columbia University Press, New York, USA.

De Souza, H.M.L., A.B. Da Cunha and E.P. Dos Santos. 1970. Adaptive polymorphism of behavior evolved in a laboratory population of *Drosophila melanogaster*. American Naturalist 104: 175-189.

Ehrlich, P.R. and P.H. Raven. 1964. Butterflies and plants: a study in coevolution. Evolution 18: 586-608.

Foster, J.E., H.W. Ohm, F.L. Patterson and P.L. Taylor. 1991. Effectiveness of deploying single-gene resistance for controlling damage by the Hessian fly (Diptera: Cecidomyiidae). Environmental Entomology 20: 964-969.

Futuyma, D.J. 1970. Variation in genetic response to interspecific competition in laboratory populations of *Drosophila*. American Naturalist 104: 239-252.

Futuyma, D.J. and M. Slatkin (eds). 1983. Coevolution. Sinauer, Sunderland, USA.

Galil, J. and D. Eisikowitch. 1968. On the pollination ecology of Ficus religiosa in Israel. Phytomorphology 18: 356-363.

Galil, J. and D. Eisikowitch. 1970. Studies on mutualistic symbiosis betweem syconia and sycophilous wasps in monoecious figs. New Phytology 70: 733-787.

Gallun, R.L. 1977. Genetic basis of Hessian fly epidemics. Annals of the New York Academy of Science 287: 223-229.

Goldschmidt, R. 1933. Some aspects of evolution. Science 78: 539-547.

Gottard, K. and S. Nylin. 1995. Adaptive plasticity and plasticity of adaptation: a selective review of plasticity in animal morphology and life history. Oikos 74: 3-17.

Gould, S.J. and R.C. Lewontin. 1979. The spandrels of san Marco and the panglossian paradigm: a critique of the adaptationist programme. Proceedings of the Royal Society, London B 205: 581-598.

Huxley, J. 1942. Evolution, the Modern Synthesis. George Allen & Unwin, London, UK.

Inbar, M. and D. Wool. 1995. Phloem-feeding specialists sharing a host tree: resource partitioning minimizes interference competition among galling aphid species. Oikos 73: 109-119.

Kammerer, P., 1924. The Inheritance of Acquired Characteristics. Boni and Liveright Publ., New York, USA.

Krimbas, C.B. 1984. On adaptation, Neo-Darwinian tautology and population fitness. Evolutionary Biology 5: 1-57.

Lewontin, R.C. 1965. Selection for colonizing ability. pp. 77-91. In: H.G. Baker and G.L. Stebbins (eds.), Genetics of Colonizing Species, Academic Press, New York, USA.

Masaki, S. 1978. Climatic adaptation and species status in the lawn cricket. II. Body size. Oecologia 35: 343-356.

Mayr, E. 1942. Systematics and the Origin of Species. Columbia University Press, New York, USA.

Morgan, T.H. 1925. Evolution and Genetics. Princeton University Press, Princeton, USA.

Nevo, E. 1979. Adaptive convergence and divergence of subterranean mammals. Annual Review of Ecology and Systematics 10: 269-308.

Ray, C. 1960. The application of Bergmann's and Allen's Rules to the poekilotherms. Journal of Morphology 106: 85-108.

Robertson, F.W. 1965. The analysis and interpretation of population differences. pp. 95-113. In: H.G.Baker and G.L. Stebbins (eds.), Genetics of Colonizing Species, Academic Press, New York, USA.

Schmidt-Nielsen, K. 1964. Desert Animals: Physiological problems of heat and water. Oxford, UK.

Scriber, J.M. 1983. Evolution of feeding specialization, physiological efficiency, and host races in selected Papilionidae and Saturnidae. pp. 373-412. In: R.F. Denno and M.S. McClure (eds.), Variable Plants and Herbivores in Natural and Managed Systems, Academic Press, New York, USA.

Seaton, A.P.C. and J. Antonovics. 1967. Population inter-relationships. I. Evolution in mixtures of Drosophila mutants. Heredity 22: 19-35.

Shkolnik, A., K. Maltz and I. Choshniak, 1979. The role of ruminant's digestive tract as a water reservoir. pp. 731-742. In: Y. Ruckebusch and P. Thievend (eds.), Digestive Physiology and Metabolism in Ruminants. MTP Press, USA.

Slobodkin, L.B. 1968. Toward a predictive theory of evolution. pp.187-205. In: R. Lewontin (ed.), Population Biology and Evolution, Syracuse University Press, USA.

Sokal, R.R., E.H. Bryant and D. Wool. 1970. Selection for changes in genetic facilitation: negative results in Tribolium and Musca. Heredity 25: 299-306.

Stone, G.N. and P. Sunnucks. 1993. Genetic consequences of an invasion through a patchy environment: the cynipid gall wasp Andricus quercuscalicis (Hymenoptera: Cynipidae). Molecular Ecology 2: 251-268.

Sullivan, R.L. and R.R. Sokal. 1963. The effects of larval density on several strains of the housefly. Ecology 44: 120-130.

Thompson, J.N. and O. Pellmyr. 1992. Mutualism with pollinating seed parasites amid co-pollinators: constraints on specialization. Ecology 73: 1780-1791.

Van Valen, L. 1973. A new evolutionary law. Evolutionary Theory 1: 1-30.

Wallace, B. 1984. Adaptation, Neo-Darwinian tautology and population fitness: a reply. Evolutionary Biology 5: 59-71.

Wiebes, J.T. 1979. Co-evolution of figs and their pollinators. Annual Review of Ecology and Systematics 10: 1-3.

Wool, D. 1973. Size, productivity, age and competition in Tribolium populations subjected to long-term serial transfer. Journal of Animal Ecology 42: 183-200.

Wool, D. and E. Sverdlov. 1976. Sib-mating populations in an unpredictable environment: effects on components of fitness. Evolution 30: 119-129.

Wool, D. 1977. Genetic and environmental components of morphological variation in gall-forming aphids (Homoptera, Aphididae, Fordinae) in relation to climate. Journal of Animal Ecology 46: 875-889.

Wool, D., R. Aloni, O. Ben Zvi and M. Wollberg. 1999. A galling aphid furnishes its home with a built-in pipeline to the host food supply. Entomologia Experimentalis et Applicata 91: 183-186.

Wool, D. 2004. Galling aphids: specialization, biological complexity, and variation. Annual Review of Entomology 49: 175-192.

Yom Tov, Y., W.O. Green and J.D. Coleman. 1986. Morphological trends in the common brushtail possum, Trichosurus vulpecula, in New Zealand. Journal of Zoology 208: 583-593.

Ziv, Y. and J.L. Bronstein, J.L. 1996. Infertile seeds of Yucca schottii: a beneficial role for the plant in the yucca-yucca moth mutualism? Evolutionary Ecology 10: 63-76.

CHAPTER 14

Natural Selection and Polymorphism

Natural selection can bring about the replacement of one genotype (or rather, one phenotype) by another, as in the case of industrial melanism and insecticide resistance. It can also maintain a balanced polymorphism in the population.

DEFINITIONS

When a population is composed, simultaneously, of two or more discrete phenotypic classes – with no intermediate forms, it is referred to as polymorphic.

Cases of discrete phenotypic classes usually have a simple genetic basis. Examples are the different eye-color mutations in *Drosophila*, body and eye-color mutants in the flour beetle *Tribolium*, or the A, B, O blood-group variation in humans. For a variable population to be called polymorphic there is a further requirement, that the frequency of the rarest class be higher than explainable by recurrent mutations (a low limit of 5% is often stated).

Many of the earliest reports on polymorphism came from morphological studies (e.g. in Ford, 1964) or chromosomal studies of *Drosophila*. In the 1960s – 1980s, many more populations were described as polymorphic in electrophoretic studies of many species (reviews in Nevo, 1978; Lewontin, 1974; Powell, 1975).

When present, a polymorphism may be **stable** – when the frequencies of two (or more) phenotypes remain unchanged at least for the duration of the observation period. Or it may be **transient**. A transient polymorphism may seem stable if the selective elimination of one morph is slow: both forms will still appear in consecutive samples, giving the impression of co-existence. It may be difficult to identify the polymorphism as transient if the intensity of the sampling efforts is low, and the length of the observation period short, relative to the speed of the selective process (Fig.14.1).

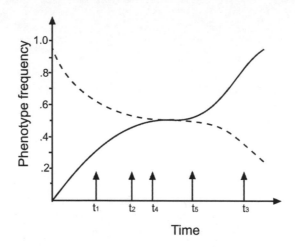

Fig. 14.1 Stable and transient polymorphism. Sampling at times t4 and t5 give the impression that the polymorphism is stable, but sampling at times t1, t2 and t3 show that one phenotype is decreasing and the other increasing in frequency.

The process most commonly offered to explain the maintenance of stable polymorphism is that of balancing selection. A polymorphism maintained by balancing selection is referred to as a **balanced polymorphism**.

Balanced polymorphisms may result from 1) selection favoring heterozygotes (see Chapter 9), 2) frequency-dependent selection, or 3) environmental heterogeneity in space or in time. Examples of all three processes follow.

SELECTIVE ADVANTAGE TO HETEROZYGOTES

If heterozygotes are favored by selection, all genotypes are expected to appear in the population every generation. Even when the homozygotes are lethal, they will continue to appear by segregation from the heterozygotes. This explanation is the most often used to explain cases of balanced polymorphism.

It is, however, somewhat disappointing to observe that in textbooks (this one is no exception) and papers on polymorphism, the only example brought forward to demonstrate balanced polymorphism by heterozygote advantage is the case of sickle-cell anemia (Hb^s) polymorphism in humans. Surely if this were a common mechanism in operation in nature, many more convincing examples should have been available.

Suppose we find a balanced polymorphism in nature and obtain the frequencies of the alleles at the marker locus at or near the equilibrium point. Theory tells us that at equilibrium, the allele frequencies are only a function of the ratio of the selection coefficients (see Chapter 9) –

$$q = s / (s + t)$$

If we can estimate the fitness disadvantage (selection coefficient) of one of the homozygotes, say **s**, and calculate the other selection coefficient (**t**) from the equilibrium formula, we may test the validity of the explanation. Merrell (1969) showed in this manner that in many situations, heterozygote advantage cannot be a valid explanation of the observed polymorphism – particularly if **q** is low. For example, if **q** < 0.2, and one of the homozygotes is very fit (1 – **s** = 0.9 or 0.8) the other homozygote must be extremely unfit, with 1 – **t** close to zero. Often a simple observation may show that this is not the case.

FREQUENCY-DEPENDENT SELECTION: RARE-MALE ADVANTAGE

In the 1960s, population geneticists were greatly interested in the report that in *Drosophila*, mating success was a function of genotype frequencies. Mutant (white-eyed) and wild-type flies were placed in the mating chamber in unequal proportions. When rare, the mutant males mated more often than predicted from their frequency in the chamber. When the frequency of the mutants increased and they became the majority, the relative advantage was reversed (Petit and Ehrman, 1969; Ehrman, 1969). The criterion for mating success was

$$k = (A / a)/(B / b)$$

where A and B were the frequencies of females fertilized by two competing male genotypes (say **ww** and **++**) and **a** and **b** were the numbers of the two male types in the mating chamber (the sex ratio in the mating experiments was kept constant at 1).

This report was interesting because frequency-dependent selection could bring about a balanced polymorphism. Male advantage, when rare, will increase the frequency of the rarer allele. As the frequency of the rare type increases, the advantage diminishes until an equilibrium point is reached. When the rare allele becomes common, its advantage is reversed, allowing the other competitor to increase. Oscillations of this kind

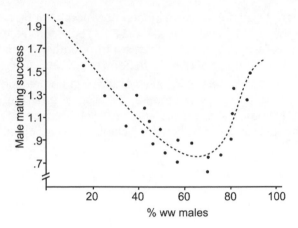

Fig. 14.2 Frequency-dependent selection: the mating success of **ww** males when placed in mating chambers in different proportion to ++ males (sex ratio maintained at 1:1). After Petit and Ehrman (1969).

will create a stable equilibrium whenever the frequencies are accidentally disturbed (Fig.14.2).

Efforts to demonstrate rare-male advantage continued for years (Spiess, 1977). Flies carrying chromosomal inversion markers, which were not supposed to have deleterious effects on fitness like the eye-color mutants, were used in many experiments. In one series of experiments, 25 pairs of flies of two strains were introduced to the observation chamber in the ratios 1:4, 1:1, and 4:1 (the two competing fly strains were differently marked because the inversion markers are not visible in the adult fly). The males of the minority strain mated more often than expected. The phenomenon was repeated with different strains and species of *Drosophila*.

How did the flies recognize which was the rare strain? In a carefully-designed experiment, visual and tactile stimuli were first eliminated. Odors remained as the only source of this information. Many flies of the rare strain were then held in a nearby cage, connected by a tube with the observation chamber. When a gentle stream of air was blown from the cage to the observation chamber, the rare-male advantage was not observed. The minority males had an advantage when the air was blowing in the reverse direction. The flies could apparently sense (by the intensity of the odor?) how frequent each strain was.

Rare-male advantage has rarely been reported in organisms other than *Drosophila* – perhaps because testing for it requires laboratory populations that can be easily-handled, and few organisms are available for this kind of research. For a theoretical review consult O'Donald (1980).

A process of frequency-dependent selection can occur also in ecological situations. When a predator preferentially removes the most abundant morph of prey, it gives an advantage to the less-common morphs (Clarke, 1979). In cases of mimicry, there will be an advantage to the mimic as long as the model is scarce: when the mimic becomes prevalent, it will lose its advantage (and the model may suffer as well) (Clarke, 1979; Ayala and Campbell, 1974).

ENVIRONMENTAL HETEROGENEITY

If the environment is heterogeneous, either in space (different sites present different requirements) or in time (seasonal climatic changes, for example), different selection pressures may operate and different morphs may be favored at different sites or times. A stable polymorphism may then be maintained. This kind of polymorphism is called **multiple-niche polymorphism** (Levene, 1953; Lewontin et al., 1978).

That the environment presents different requirements in space or time can be discovered rather easily by measurements, or by observation of habitat preferences of some organisms. Environmental heterogeneity was invoked to explain many cases of electrophoretic polymorphism. Also, in many studies of the shell-color polymorphism of the European land snail, *Cepaea nemoralis*, no other mechanism seemed to explain the stability of morph frequencies in different sites over a period of several years (Jones et al., 1977). Selective predation of shells by birds was unlikely: the snails were not a preferred food of the thrushes, and were consumed only when other food was limited. Climatic differences among sites possibly affected shell color frequencies – perhaps due to differences in heat absorbance in dark versus light-colored shells. But this was not always convincing.

Random effects could also be important. Many populations of the snail are relatively small and isolated – among other reasons, by the destruction of the habitat – and the phenotype frequencies could be affected by random drift. Moreover, the reproductive rate of the snails in some years is very low, creating genetic bottlenecks. Investigations of *Cepaea* in the USA indicate that they were founded by small numbers of immigrants from Europe. 'Founder effects' could also be a factor in British populations, where much of the work on the snails was carried out. None of these mechanisms is by itself sufficient to account for all the known facts, and any of them – or combinations of them – could be effective in individual cases (Hedrick et al., 1976; Jones et al., 1977).

POLYMORPHISM WITHOUT SELECTION

Polymorphism may be maintained in populations by mechanisms not involving natural selection at all.

Sexual dimorphism is a widespread phenomenon. In many animal species, males and females are different in size and color, a fact of which Darwin was well aware. The sexes coexist of course in the same populations in a balanced ratio – a situation which deserves to be labeled as a balanced dimorphism.

This dimorphism is determined, in many organisms, by the different genetic content of the two sexes. In humans and *Drosophila* one chromosome – x – accounts for the difference: the female is homogametic (xx) and the male is heterogametic (xy). The reverse is true in butterflies, and in aphids and other bugs the male lacks one chromosome (x0). In bees, wasps and ants, the males develop from unfertilized eggs and have half the genetic material that the females carry. Some of these chromosomal differences must dictate, directly or indirectly, the morphological differences between the sexes.

Sexual Selection: Historical Notes

Darwin (in 'Sexual Selection', a major part of his book "The Descent of Man" (1874)), sought to explain the evolution of the different male and female phenotypes. He saw no way that natural selection could be the cause of differences in 'secondary sexual characters' – like color patterns – between males and females, characters which seemed to contribute nothing to the survival and reproduction (fitness) of the organisms. Darwin suggested that these characters evolved by **sexual selection.** This mechanism was supposed to work not by competition of organisms for limited resources, but by competition of male against male within populations for the attraction or possession of the females. According to Darwin, it is the female who chooses her mate, not the other way around. The reason is that a male can fertilize several females, and his interest should be to increase the number of mates as much as possible. The female, on the other hand, is limited by her fecundity and the need to care for the brood, so she should be interested in brood quality, and should be careful in her choice of a mate.

The question is, how can a female tell which male is the best? Darwin believed that she uses the 'secondary sexual characters' as clues. For example, in birds, colors, ornamental structures like the peacock's tail, and song – enable the female to choose the best of all available males as father of her offspring. This sexual selective pressure led to the improvement – sometime in an extravagant manner – in color pattern or song in the males:

Does the male parade his charms with so much pomp and rivalry for no purpose? Are we not justified in believing that the female exerts a choice, and receives the addresses of the male who pleases her most?

We can perceive that the males [of birds] , which during former times were decked in the most elegant and novel manner, would have gained an advantage, not in ordinary struggle for life, but in rivalry with other males, and would have left a larger number of offspring to inherit their acquired beauty (Darwin, 1874, p.769).

Interestingly, A.R. Wallace – Darwin's co-discoverer of natural selection, whose ideas are often reported in 'The Descent of Man' – disagreed with Darwin's interpretation of the evolution of sexual dimorphism. Being a field naturalist, familiar with the breeding habits of animals in nature, Wallace suggested that it is not the male bird's bright coloration and distinct song which were favored by selection, but rather the dull, camouflaged color of the female was selected to give her protection from predators while sitting on her eggs. Brightly colored females are conspicuous against the background, and run the risk of being taken by predators. Thus the female and young do not have bright colors – while the male is free to vary. As a proof, Wallace points out that in bird species that nest in holes and other protected sites, like kingfishers and bee-eaters, both sexes are very colorful.

Darwin reports Wallace's opinion, but sticks to his own explanation of the evolution of sexual dimorphism by sexual selection. According to Darwin, sexual selection works differently in birds and in mammals: male birds compete by showing off their plumage or song. In mammals, brute force in battle is the selective force between males, and sexual selection consequently led to an increase in male size and the development of structures used in battle, like horns and the canine teeth.

Sexual dimorphism exists in many other vertebrate and invertebrate species, from crabs and insects (butterflies and beetles in particular) to fish, amphibians, and reptiles. Darwin explains their evolution by sexual selection. Some of his arguments are less convincing than Wallace's, who sticks to natural selection.

Darwin's theory of sexual selection was severely criticized by T.H. Morgan, and rejected as an explanation of sexual dimorphism (Morgan, 1903).

A Recent Example: Dimorphism in Horn Size in Scarabaeid beetles

That scarabaeid beetles – including the genus *Onthophagus* – have horns was well known to Darwin, who found it difficult to explain their evolution by his sexual selection theory:

The extraordinary size of the horns, and their widely different structures in closely allied forms, indicate that they have been formed for some purpose. But their excessive variability in the males of the same species leads to the inference that this purpose cannot be of a definite nature…

The most obvious conjecture is that they are used by the males for fighting together. But the males have never been observed to fight…The conclusion that the horns have been acquired as an ornament is that which best agrees with the fact of their having been so immensely, yet not fixedly, developed (Darwin, 1874, p.644).

A case of facultative dimorphism in horn size was recently described in a series of papers on a small (3-6 mm long) scarabaeid beetle, *Onthophagus taurus* (Moczek and Emlen, 2000). *O. taurus* inhabits dung pats in the field in North Africa and Asia, and was introduced accidentally into the northern USA in the 1970s (Moczek, 1998). The females dig tunnels under the dung and drag in pieces of dung from which they mold brood balls. A single egg is laid in each ball, on which the larva feeds until pupation.

Males of this beetle are of two morphs: large males have long frontal horns, and small males (and females) have no horns (see Fig. 10.11 above). Recent research showed that at least as regards *O. taurus*, Darwin's 'obvious conjecture' was correct: the large males do use their horns for fighting together, and for dislodging competitors from their breeding tunnels (Moczek and Emlen, 2000). Large males guarded the tunnels where the females were preparing the brood balls, and engaged in head-to-head fights to gain possession of the females. The fights ended when one of the contestants left the tunnel. The smaller, hornless males did not engage in fights: instead, they used 'sneaking' tactics to bypass the guarding males and mate with the females. The females were never observed to fight nor to reject a courting male, and did not prefer horned to hornless males. The dimorphism in presence and size of horns has no genetic component, however, and longer horns could not have been selected – either by sexual or natural selection: the character is totally determined by the quantity and quality of the food the individual larva finds packed for it by its mother. The presence of horns was related to beetle size (weight) but not linearly: the males had to be of a threshold critical size in order to develop these cuticular structures. Regression analysis of body size and of horn length of the offspring on the parental characters showed that the variation in these characters was entirely environmental, with no genetic component.

[Although Moczek and Emlen repeatedly state that horn length and beetle size were not heritable, they suggest that during a long history of selection [diversifying selection?] males were selected to adopt one of the two mating strategies – fighting or sneaking – and intermediates were selected against. An environmentally-controlled switch mechanism in ontogeny evolved to determine the strategy for each adult by the food supply, resulting in the observed dimorphism.]

Hormonally-controlled Polymorphism

In some organisms, sex and sexual dimorphism are hormonally controlled, activated by an environmental stimulus. In many fishes, the young are morphologically female, and the change of sex to male – including a change in morphology, coloration and behavior – depends on school density and the presence or absence of other males (e.g. Fishelson, 1975.

In social insects like termites and bees, there are well-known examples of hormonal – social control of caste polymorphism. In termites, the balance between reproductive and worker morphs in the colony is maintained by pheromones produced by the sexual female and male, which prevent the development of gonads in the larvae. In social bees, this is affected by the workers giving special attention to certain larvae destined to become queens, and feeding them with special food. The males develop from unfertilized eggs. In these cases natural selection within colonies is not involved in the preservation of the polymorphism.

Seasonal Phenotypic Changes

In many insects, there are seasonal phenotypic changes within populations, which are correlated with – or triggered by – environmental variables. The term 'polyphenism' was suggested for such non-genetic polymorphisms (Shapiro, 1976), but it is not often used.

Good examples for this phenomenon are provided by aphids (Dixon, 1998; Wool, 2004). In the life cycle of a single species, during a sequence of parthenogenetic generations (which ensures that no genetic variation is involved), aphids may be wingless or winged (alate). Up to eight different morphs can be observed in a single lineage. All these changes are environmentally triggered. Day length, temperature and density (crowding) are known to induce the transition between wingless and winged morphs (Dixon, 1998).

Seasonal changes in morphology were reported in butterflies. The quantity of melanin (black) and pteridin (orange) – two components of wing color in the American butterfly *Colias eurytheme* – varied seasonally in relation to thermoregulation (Shapiro, 1976). The color of the pupa – green or brown – in species of papilionid butterflies within the same species, seems to be determined by the texture and the color of the substrate on which the pupating larvae are moving (Hazel and West, 1979).

POLYMORPHISM IN THE LADYBIRD BEETLE, *ADALIA BIPUNCTATA*

The well-studied color polymorphism in *Adalia bipunctata* may illustrate the difficulties of interpreting the observed data on polymorphic populations.

The lady beetle, *A. bipunctata*, is widespread in many areas in Europe and often occurs in large numbers. The typical color of the elytra is red with two black spots (hence the scientific name). In some places – in particular in industrial areas – a large proportion of the beetles are black (melanic). This attracted the attention of investigators, who hoped to find another example of polymorphism maintained by selection, like the case of industrial melanism in moths.

The genetic control of the polymorphism was soon shown to be complex. At least 12 alleles conferred some level of melanism on the phenotype. The main melanic alleles were dominant. The number of black spots varied from 2 to 6 (the beetles are referred to as either typical (red) or melanic, regardless of the exact phenotype). Melanics were found in high frequencies – up to 97% – in industrial areas near Glasgow in Scotland and near Manchester in central England. In a detailed investigation near a coal-refinery plant in Wales, the frequency of melanics in samples of *A. bipunctata* were correlated with distance from the plant: their proportion dropped from 55% melanics near the factory to 5 %, 17 km away, and 1% in samples at greater distances (Creed, 1974).

Many samples from various areas in England were examined together with tests of air pollution. The frequencies of melanics in these samples varied from 0 to 90% and some significant correlations were found between percentage melanics and components of smoke in the air. Moreover, following air-pollution control in Birmingham in 1960-1969, the frequency of melanics in that area decreased.

Despite these facts, however, the similarity of the phenomenon to industrial melanism in moths is no proof that the same mechanism maintains the polymorphism of *A. bipunctata*. Selective predation by birds does not seem likely in lady beetles. When disturbed, the beetles exude drops of yellow liquid from the bases of their legs, and they are usually rejected by captive birds when offered to them. Frozen beetles were offered to hungry captive birds; some were eaten, but regardless of color. Further, both red and black are equally visible on the green or brown background of grass or leaves, and neither can be considered protective (red and black are usually regarded as warning coloration).

Muggleton (1978) argued that the behavior of the predators in nature may be very different from that observed in cages. He recorded 121 anecdotal reports of birds in nature eating the lady-beetles versus only 19 cases in which they were rejected. In a controlled study in nature with frozen beetles arranged in a 'live' position on leaves, 28% were eaten, and a pair of swallows fed their young with the beetles throughout the breeding season.

O'Donald and Muggleton (1979) often found an excess of melanic individuals among copulating pairs of the beetles than in randomly-collected field samples. This suggested that the beetles were not mating at random (Muggleton, 1979). The suggestion, however, was not supported statistically.

Melanic beetles could have a physiological advantage in cold climates because they may warm up faster than typicals and begin activity earlier in the day or the season. In a laboratory experiment, melanics warmed up more quickly in sunlight compared with typical beetles (Brakefield and Willmer, 1985) – a fact that gives the melanics an advantage: they hatch earlier from the pupae, emerge earlier from dormancy in the spring and start activity earlier (but they also tend to die earlier in the season. See Kearns et al., 1990).

The frequencies of melanic *Adalia* in Italy decreased with altitude, from more than 90% at sea level to 15-50% at 1000 m altitude (Scali and Creed, 1975) – and this trend was attributed to the decrease in air pollution (most sources of air pollution, such as cities and industrial plants, are located in coastal areas) or to differences in mean annual temperatures. In Norway, a positive correlation was reported between melanic frequencies of *Adalia* and climatic factors like temperature, rainfall, and distance from the sea.

Adalia bipunctata in the Netherlands hibernates in large numbers in crevices in trees and buildings during the winter (Brakefield, 1984a,b). Climatic factors – particularly temperature – were correlated with the geographic distribution of melanic frequencies.

The difficulties in interpreting the data on melanic polymorphism in *A. bipunctata* still remain after 20 years or more of research. This is an example of the difficulty in interpreting any phenomenon in nature, where we try to deduce the mechanism from the end result of a long-term process.

Literature Cited

Ayala, F.J. and C.A. Campbell. 1974. Frequency-dependent selection. Annual Review of Ecology and Systematics, 5: 115-138.

Brakefield, P.M. 1984a. Selection along clines in the ladybird *Adalia bipunctata* in the Netherlands: a general mating advantage to melanics and its consequences. Heredity 53: 37-49.

Brakefield, P.M. 1984b. Ecological studies on the polymorphic ladybird *Adalia bipunctata* in the Netherlands. II. Population dynamics, differential timing of reproduction and thermal melanism. Journal of Animal Ecology, 53: 775-790.

Brakefield, P.M. and P.G. Willmer. 1985. The basis for thermal melanism in the ladybird *Adalia bipunctata*: differences in reflectance and thermal properties between the morphs. Heredity 54: 9-14.

Clarke, B.C. 1979. The evolution of genetic diversity. Proceedings of the Royal Society, London, B 205: 453-474.

Creed, E.R. 1974. Two-spot ladybirds as indicators of intense air pollution. Nature 249: 390-392.

Darwin, C. 1874. The Descent of Man. 2nd ed. Hurst & Co., New York, USA.

Dixon, A.F.G. 1998. Aphid Ecology: an optimization approach. Prentice-Hall. London U.K.

Ehrman, L. 1969. The sensory basis of mate selection in *Drosophila*. Evolution 23:59-64.

Fishelson, L. 1975. Ecology and physiology of sex reversal in *Anthias squamipennis* (Peters), (Teleostei: Anthiidae). pp. 283-294. In: R. Reinoth (ed.), Intersexuality in the Animal World, Springer, New York, USA.

Ford, E.B. 1964. Ecological Genetics. Methuen, London, UK.

Hazel, W.N. and D.A. West. 1979. Environment control of pupa colour in swallowtail butterflies (Lepidoptera: Papilionidae): *Battus philenor* (L.) and *Papilio polyxenes* Fabr. Ecological Entomology 4: 393-400.

Hedrick, P.W., M.E. Ginevan and E.P. Ewing. 1976. Genetic polymorphism in heterogeneous environments. Annual Review of Ecology and Systematics 7: 1-32.

Jones, J.S., B.H. Leith and P. Rawlings. 1977. Polymorphism in *Cepaea*. Annual Review of Ecology and Systematics 8: 109-144.

Kearns, P.W.E., I.P.M. Tomlinson, P. O'Donald and C.J. Veltman. 1990. Non-random mating in the two-spot ladybird (*Adalia bipunctata*). I. Reassessment of the evidence. Heredity 95: 229-240.

Levene, H. 1953. Genetic equilibrium when more than one ecological niche is available. American Naturalist, 87: 331-333.

Lewontin, R.C. 1974. The Genetic Basis of Evolutionary Change. Columbia University Press, New York, USA.

Lewontin, R.C., L.R.ginzburg and S.D. Tuljapurkar. 1978. Heterosis as an explanation for large amounts of genic polymorphism. Genetics 88: 149-170.

Merrell, D.J. 1969. Limits on heterozygous advantage as an explanation of polymorphism. Journal of Heredity 60: 180-182.

Moczek, A.P. 1998. Horn polyphenism in the beetle *Onthophagus taurus*: larval diet quality and plasticity in parental investment determine adult body size and male horn morphology. Behavioral Ecology 9: 636-641.

Moczek, A.P. and D.J. Emlen. 1999. Proximate determination of male horn dimorphism in the beetle *Onthophagus taurus* (Scarabaeidae). Journal of Evolutionary Biology 12: 27-37.

Moczek, A.P. and D.J. Emlen. 2000. Male horn dimorphism in the scarab beetle *Onthophagus taurus*: do alternative reproductive tactics favor alternative phenotypes? Animal Behavior 59: 459-466.

Morgan, T.H. 1903. Evolution and Adaptation. MacMillan, New York, USA.

Muggleton, J. 1978. Selection against the melanic morphs of *Adalia bipunctata* (two-spot ladybird): a review and some new data. Heredity 40: 269-280.

Muggleton, J. 1979. Non-random mating in wild populations of polymorphic *Adalia bipunctata*. Heredity 42: 57-65.

Nevo, E. 1978. Genetic variation in natural populations: pattern and theory. Theoretical Population Biology 13: 121-177.

O'Donald, P. and J. Muggleton. 1979. Melanic polymorphism in ladybirds maintained by sexual selection. Heredity 43: 143-146.

O'Donald, P. 1980. A general analysis of genetic models with frequency-dependent mating. Heredity 44: 309-320.

Petit, C. and L. Ehrman. 1969. Sexual selection in *Drosophila*. Evolutionary Biology 3: 177-223.

Powell, J.R. 1975. Protein variation in natural populations of animals. Evolutionary Biology 8: 79-119.

Scali, V. and E.R. Creed. 1975. The influence of climate on melanism in the two-spot ladybird, *Adalia bipunctata*, in central Italy. Transactions of the Royal Entomological Society, London 127: 163-169.

Shapiro, A.M. 1976. Seasonal polyphenism. Evolutionary Biology 9: 259-333.

Spiess, E.B. 1977. Genes in Populations. Wiley, New York, USA.

Wool, D. 2004. Galling aphids: specialization, complexity, and variation. Annual Review of Entomology 49: 175-192.

CHAPTER 15

Classification of Selection Processes

In most cases it is impossible to measure the selection process as it occurs, but it is often possible to learn about the direction of the process from considerations of the observable results.

Three kinds of selection processes can be described: stabilizing selection, directional selection and diversifying selection. This classification dates back to the 1950s, when it was applied in experiments on quantitative characters. The selection processes can be illustrated by their effects on the shapes of the frequency distributions of the selected traits (Fig. 15.1)

Stabilizing selection works against the phenotypes distant from the mean on either side (Fig. 15.1A). The result is a reduction in the phenotypic

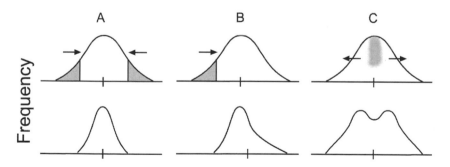

Fig. 15.1 The effects of selection pressure on the frequency distributions of the selected traits. Top: character distribution before selection. Arrows indicate the direction of selection. Bottom: distribution after selection. A: balancing selection. B: directional selection. C: diversifying selection.

variance so that the population after selection is distributed more closely around the mean than before selection. In a normally-distributed trait, the resulting distribution is described statistically as leptokurtic.

When the environment remains unchanged for a long time, a single or a small number of genotypes may perhaps perform best while the deviants from the mean are less successful. An example of a character under stabilizing selection may be the weight (size) of human babies at birth: those born at a very low weight have a smaller chance of surviving; while babies much larger than the average may endanger the lives of their mothers. Both of these sources of selective mortality have been reduced by modern medical care, but were effective 50 or 100 years ago, before the introduction of incubators and safe caesarian sections.

Drosophila genotypes with an average number of abdominal setae – a well-investigated quantitative trait – were better fit than individuals with larger or smaller numbers of setae.

Directional selection. When individuals larger (or smaller) than the mean are selected against, an asymmetrical shift in the mean of the distribution results (Fig. 15.1 B). For example, selection for an increase (as well as for a decrease) in the number of abdominal setae was effective in laboratory populations of *Drosophila*. Directional selection for large size may occur in nature, for instance, if predators consistently and preferentially remove smaller prey, or if larger individuals have a better chance of procuring food. The majority of selection pressures applied by man to domesticated plants and animals are directional.

Disruptive selection is more difficult to demonstrate. This is a process where individuals with the character value close to the population mean are selected against, favoring the deviants higher and lower than the mean (but without allowing the selected ends to mate together) (Fig. 15.1C). In nature, if two food plants are available, different genotypes may specialize on the alternative hosts while the intermediate individuals suffer a lower fitness. Such a scenario may result in the formation of two host-races and perhaps lead to sympatric speciation (Chapter 21). Complex experimental designs have been employed (with varying success) to demonstrate that such a process can be effective (e.g., Alicchio and Palenzona, 1974).

WHAT CAN NATURAL SELECTION DETECT?

In the 1900s, the population geneticist J.B.S. Haldane believed that natural selection can detect single-allele substitutions. He based this belief on the observation of industrial melanism, where an allele specifying one phenotype was substituted by another, specifying a protected phenotype.

Similar single-allele substitutions are detectable in many cases of insecticide resistance in insects (review by Roush and Mackenzie, 1987).

However, Lewontin (1970, 1974) and other investigators suggested that selection works with gene complexes, not isolated genes. Selection may lead to great morphological changes and even reproductive isolation – without detectable changes in chromosomal or electrophoretic marker genes. As genes are associated in linkage groups, changes in marker genes may be caused by selection on other genes (Franklin and Lewontin, 1970). Genetic markers located on different chromosomes may show a similar pattern of departure from expectation ('linkage dis-equilibrium'). Natural selection can change the frequency of a 'neutral' gene if it is associated with other, selected genes (e.g., the black allele in *Tribolium*. Stam, 1975).

Although it is mathematically convenient to work with single (marker) loci, this is an over-simplification when we deal with natural populations. Selection operates on entire individual phenotypes. The result of selection is the proliferation or removal of individuals – entire genomes – although our only way to detect the fact is to concentrate on individual gene markers.

EVOLUTION OF ALTRUISTIC TRAITS

Many evolutionists found it difficult to explain how behavioral traits that seem to be deleterious for the individual performing them, could evolve by natural selection. For example, when a group of hyraxes, ground squirrels or meerkats, come out from their hiding places to feed, one of them is always on the lookout to warn its fellows of danger. This individual loses feeding time while on duty, and runs the risk of attracting the predator to itself. Such an 'altruistic' behavior was investigated in babblers (*Turdoides*, Timaliidae), social birds that inhabit the Arava desert valley in Israel. A higher-ranking individual in the group of babblers often feeds some individuals of lower rank with food items it has caught, instead of eating them itself – as if making an effort to help his subordinates. Moreover, in a group of babblers, only one pair breeds – and all other individuals forsake their reproductive efforts and help the pair in feeding their brood! (Zahavi and Zahavi, 1996). In social insects – like bees and wasps – such seemingly 'altruistic' behaviors are common. The most famous examples are the colonies of the honeybee and termites, where only one female lays fertile eggs, and thousands of other individuals build the nest and feed the queen and the brood. A less advanced system is found in *Polistes* paper wasps: several fertile females ('queens') share the same comb, but one of them lays the majority of the eggs, while the others help in building the comb and feeding the young.

These behaviors are beneficial (actually vital) for the colony, but seem to be contrary to the good (or interest) of individuals: each individual should try to maximize its own reproductive potential and propagate its own genotype. The evolution of 'altruistic' traits needs explanation: behaviors which are harmful to the individual should have been removed long ago by natural selection.

The concept of <u>altruism</u> stems from observations of human behavior. Wars between tribes and nations – a characteristic of human history from its early beginnings – are conceived as cases of conflict of interest, where a man is called up to fight (and possibly die) for his country, sacrificing his own life and reproductive future for the good of others. There are other examples of altruistic behavior in humans, like risking one's life to save another from drowning in the sea or from a burning building etc (a grand example was the behavior of the fire-fighters and policemen who entered the collapsing World Trade Towers in New York on September 11, 2001). I shall not discuss the motivation behind such human behavior. But it is doubtful that animals – insects in particular – are motivated by noble considerations of helping others.

ALTRUISM AND THE HANDICAP PRINCIPLE

The evolution of another kind of trait is also difficult to explain by natural selection – the extreme development of ornaments. The peacock's tail, or the antlers of the now-extinct Irish elk, may have been acquired because they were advantageous to their carriers in competition between individuals within species – by natural as well as sexual selection. But most of these over-specializations led to extinction, like the huge body size of the Miocene dinosaurs: it seems that in the long run, they were disadvantageous to the species as a whole.

Detailed studies of the babblers (*Turdoides*) in the Arava valley of Israel, over more than 30 years (Zahavi and Zahavi, 1996), indicated that the seemingly 'altruistic' behavior of the birds towards each other – are not without benefit to the performer. These behaviors serve to advertise the strength and quality of the performer and to maintain its status in the social group. Zahavi (1975) suggested a '**handicap principle**' to account for the evolution of characters or behaviors which are harmful to the carrier – like the peacock's tail or the antlers of the Irish elk. The handicap theory states that females cannot know the quality of potential mates, and may be using the exaggerated characters as signals to estimate that quality. Characters that are handicaps for the male, or even endanger his life, become a reliable signal of his quality: if he has survived despite carrying this burden, he must have some other positive characters as well, and should be a good potential mate, able to pass on these good genes to his offspring.

For example, territorial male birds tend to occupy as large a territory as they can early on in the season. This requires time and energy expenditure for defending the large territory against intruders, instead of feeding (= a handicap). But a large territory provides a large area in which to hunt for food, and a female choosing a male which defends a large territory, has a good chance of being able to provide for a large brood.

In mammals, unlike birds, Darwin wrote in his book Sexual Selection, "the male appears to win the female much more through the law of battle [winner takes all] than the display of his charms" (Darwin, 1874, p.818). But he considered that female choice plays a role too:

> It would be a strange anomaly if female quadrupeds ... did not generally, or at least often, exert some choice... The suspicion has crossed my mind that [the branching horns of deer] may serve in part as ornaments... that they are ornamental in our own eyes no one would dispute (Darwin, 1874, p. 836).

These ornaments are costly to the bearers – a handicap, as illustrated by the exaggerated antlers of the extinct Irish elk. But if the antlers in deer are to serve as reliable signals, their size should be strongly correlated with male fitness. A recent study provided this kind of evidence. The size and complexity of the male antlers of red deer (*Cervus elaphus*) – when corrected for differences in body size – were correlated with qualitative and quantitative characteristics of male sperm. Female choice of the males with the largest or most complex antlers may equate the choice of the most fertile male (Malo et al., 2005).

GROUP SELECTION

Difficulties in explaining the evolution of characters deleterious to their carriers led to the suggestion that selection processes work at different levels: **individual selection** operates within populations (when individuals struggle for existence and compete with each other), and **group selection** operates on entire populations when one population has an advantage over others. These two kinds of selection may be antagonistic – group interests may involve sacrificing individual fitness, as in the case of altruistic traits (Wynne-Edwards, 1962).

> Such group-related adaptations must be attributed to the natural selection of alternative groups of individuals, and... natural selection of alternative alleles within populations will be opposed to this development (Williams, 1966, p. 92).

One scenario often cited, is the extinction and recolonization of small, isolated populations (Van Valen, 1971). If conditions for survival within populations are favorable, an emigrating individual is at risk: it is exposed to predation and may not find a suitable site in which to feed and reproduce. Individual selection should work against the tendency to disperse, and promote sedentary behavior. But a population in which a tendency to disperse is present, will have a chance of colonizing new sites if they become available. Thus group selection will promote dispersing or colonizing populations.

The processes of extinction and recolonization of local populations are the basis of **meta-population theory**. This is an ecological approach to the study of nature on large geographical scales. The concept is rather new in ecological literature, and describes the landscape – from the point of view of the organism – as a mosaic of patches of suitable habitat, embedded in a matrix of uninhabitable area (by that organism). In each patch, the organism has all the resources it requires. The connection between patches is maintained by dispersal of individuals – which also facilitates gene flow between populations and maintains the cohesion of the species. Patches may become vacant by extinction of the population, and be colonized by dispersal from other populations.

Most evolutionists oppose the idea of two levels of selection, and claim that individual selection is sufficient to account for processes both within and among populations. The controversy over whether or not group selection is a different process from individual selection was academic for many years (see Wade, 1978 for a review of the theoretical models of group selection).

Williams (1966) dedicated four chapters in his book to arguing that there are no special group-selected characters: fitness of populations cannot be measured separately from the fitness of individuals. A group character such as percentage mortality cannot be measured on individuals; nevertheless, it is nothing but the sum of the ability of individuals to withstand the source of mortality.

Group Selection: Laboratory Studies

In the 1980s, group selection was revived in a new light with the laboratory experiments of M. Wade, using flour beetles (*Tribolium*) as model organisms. Wade treated the question from a quantitative-genetic point of view, regardless of whether or not there was a conflict between the interests of the individuals and the group (Wade, 1977). With this experimental approach, group selection is detected if selection for a group character is effective (the average of a trait in each group was used as a character). The question then becomes, is there enough additive genetic variation among group means, over and above variation among individuals within groups?

The character of interest in the study was group (population) size. There were four treatments with 48 populations in each. At the start of every generation, each population was founded with a propagule size of 16 individuals, assuming a 1:1 sex ratio (Fig. 15.2).

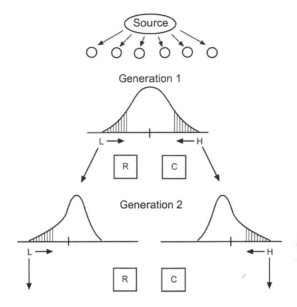

Fig. 15.2 The design of Wade's selection experiments. See text for details.

Selection for large populations (H) was imposed by ranking all populations by their output size at the end of each generation. Starting with the largest population, new propagules of 16 were set up from the available offspring; then the next largest population was taken in, until 48 new propagules were ready. Selection for small (L) populations was applied by starting at the lowest rank and going up the ranks, setting up propagules of 16 until 48 were ready. In the third treatment, populations for continuing the line every generation were picked at random (R). In the fourth treatment (control, C), one propagule was set up from the offspring of each of the 48 populations of the preceding generation (Fig. 15.2).

This experimental model simulated a case of extinction and recolonization in that, e.g. in H, an advantage in colonization was given to migrants from the most prolific populations. In L, advantage was to the least prolific ones.

The results showed that 'group' selection (as defined in this experiment) – was effective. After 8 generations, the mean population size in the H line (178 individuals per population) was higher, and in the L line it was lower (20 individuals) than in the control. There was no change in mean population size in the random treatment. The difference in mean population size persisted for three years after the selection was relaxed (Wade, 1984). In absolute numbers, relative to the initial reproductive output, population size <u>decreased</u> with time in all treatments – due to inbreeding depression? – but the decrease rate was slower in the H line and faster in the L line than in the control.

The experiment illustrates that the selected character – mean population size (the 'populational phenotype' in Wade's terminology) – is the combined reproductive output of the individuals, and not an independent attribute of the group. In the H line, individuals with a large reproductive output were more likely to be included as parents. Similarly, individuals with poor productivity were more likely to be included as parents in the L line. Had selection been applied to the individuals directly (had it been possible to recognize each individual's reproductive potential in advance) the response to selection might have been faster (see also Wool, 1982).

Craig (1982) simulated a conflict between individual and group 'interests'. The selected character was the tendency to emigrate, which is relevant to the concept of extinction and recolonization of sites in a meta-population. This trait is partially heritable in *Tribolium* (Ritte and Lavie, 1972). Emigration tendency can be measured in *Tribolium* when pairs of vials – one (source) vial containing medium and beetles, the other empty – are connected by a tube in which a string is inserted, to serve as a bridge. The string touches the medium but dangles free in the empty tube. This is a one-way emigration system: beetles which leave the source cannot return to it (and can be easily counted).

Two opposing selection pressures were applied in Craig's experiment. In selection for high (H) emigration rates, the emigrants from the five source vials with the highest percentage of emigrants were used as parents for the next generation. In the Low (L) line, the emigrants from five sources with the smallest percentage of emigrants were used. Individual selection within populations was practiced by destroying a fixed proportion of the emigrants every generation.

The result showed an increase of dispersal (emigration) tendency in H, compared with the unselected control, despite selection against dispersers within populations. (The results of Craig's experiments were probably affected by beetle density in the source vials: emigration rate was higher in more crowded replicates than in less-crowded ones.)

The criterion for 'inter-demic selection' (Wade's term), was the percentage of emigrants. A population with a large proportion of emigrating individuals must contain many individuals with strong dispersal tendency. Selection of beetles from these sources as parents of the next generation necessarily increases the difference among groups and the similarity among the individuals within groups (interestingly, in Craig's experiments, most of the inter-individual differences within groups were not genetic).

Wade and Goodnight (1991) simulated the advantages of 'inter-demic selection' by giving the most prolific population an advantage in colonizing new habitats. All populations were arrayed by their output size at the end of a generation, and their index of reproductive success was calculated relative to the average: $W_i = p_i/P$, where p_i is the output of population i and P the average of all populations. From each population that had $p_i > P$, some individuals were transferred to a 'migrant pool'. Populations in which $p_i < P$ received migrants from the pool. After a few generations the mean reproductive output increased significantly in all lines.

Inter-demic selection may be efficient if the variation among groups is large, and variation within groups is small. In cases like this, selection using mean size as a criterion will support directional selection. Again, the selected trait is the sum of individual tendencies, not an independent group trait. Inter-demic selection is individual selection in disguise.

KIN SELECTION

A model for the evolution of seemingly–altruistic behaviors in social insects was suggested by Hamilton (1964). Hamilton was specifically interested in the evolution of the system where a single female (queen) is assisted by numerous workers, which 'give up' their own reproductive potential.

In the Hymenoptera – and other haplo-diploid organisms, a female is genetically 50% similar to her daughters, but 75% similar to her sisters (Table 15.1). (In sex-linked characters, sisters have a 50% chance of inheriting the same **x** chromosome from their mother, plus 25% due to the certainty of having the same **x** chromosome from their father. In haplo-diploids, the entire genome behaves as sex-linked.) Therefore, Hamilton reasoned, an altruistic individual in the hive helping her sister to raise her brood, actually enhances the chance to propagate her own genes more than by reproducing and raising her own brood!

Hamilton coined the term **'inclusive fitness'** of an individual, incorporating the benefit from raising its brood and from helping its relatives (for 'fitness' read reproductive success). Altruistic behaviors, which are deleterious for the individual but beneficial to the kin, may be selected if the beneficial effects on fitness are greater than the cost – depending on the

Table 15.1 Genetic relatedness in a haplo-diploid system

Focal individual:		male	female
Relatedness to:	mother	1	0.5
	father	-	0.5
	sister	0.5	0.75
	brother	0.5	0.25
	daughter	1	0.5
	son	-	0.5

relatedness between the helper and its kin. An advantage for altruistic behavior will occur if

$$r_{ab} * B/C > 1$$

where r_{ab} is the relatedness of individuals **a** and **b** (Table 15.1), B is the beneficial effect (of helping **b**) to **a**'s inclusive fitness, and C is the cost to **a** from doing so.

The genetic relatedness among sisters shows that the inclusive fitness of the workers in a bee hive is greater than their expectancy from reproducing their own daughters. In some Hymenoptera, the workers lay unfertilized eggs from which males emerge. In this way they increase the proportion of their own sons (mother-son relatedness: 0.5) over that of their brothers (brother-sister relatedness: 0.25), again increasing their inclusive fitness.

Empirical Tests of Kin Selection Theory

If the kin-selection theory is correct, the frequency of altruism should increase with the genetic relatedness. Lester and Selander (1981) investigated colonies of the wasp, *Polistes*, in Texas. These wasps have some social structure, but no marked morphological differences between 'queens' and workers. Each colony is founded by one female, but other females may later join the same nest, and a hierarchy develops – based, among other factors, on the status of the ovaries. In 42 colonies of the wasps, Lester and Selander determined the ovarian status of each adult, and used electrophoretic markers to determine the relatedness of the nest mates. The investigators predicted that the females inhabiting the same nest should be sisters helping each other. According to kin-selection theory, the relatedness among them should be greater than 0.5 and close to 0.75 (Table 15.1). The prediction was not supported by the data: the relatedness between females in the same nest was often smaller than 0.5.

Altruism, kin selection and proto-sociality in insects

The essence of eusociality is a trade-off between producing one's own offspring and helping collateral kin via such activities as defense, foraging and brood rearing (Cited from Wilson, 1971, by Perry et al., 2004.)

Bees and wasps are considered 'advanced' insects, and their social structure has attracted a lot of attention and long been admired by humans. But rudiments of sociality and altruistic behavior towards colony members – mostly their kin – is reported in far less 'advanced' insects, particularly in gall-inducing thrips (Thysanoptera) and gall-inducing aphids that have a soldier caste (Aphidoidea : Homoptera).

Eusociality is normally defined by three criteria: overlap in generations between parents and offspring, cooperative brood care, and reproductive division of labor, with more-or-less sterile individuals working on behalf of fecund individuals. The common characteristic of the social thrips and aphids is that each colony lives in a gall induced by a single individual (or a single pair) and the inhabitants of the gall are genetically related.

Thrips

The gall-inducing thrips meet all three criteria, and can be considered eusocial. Six species of social thrips (of several hundred species of thrips in Australia), induce galls on species of *Acacia.* These thrips have a distinct 'soldier' morph with reduced wings and enlarged forelegs, thought to be adaptations for defense (Kranz et al., 1999, 2001).

The first cohort of the eggs of the founding pair in the gall develops into soldiers. The second batch becomes normal reproductive dispersers with fully developed wings. Although they do reproduce, the fecundity of the soldiers is lower than that of the reproductives. There is evidence that they kill kleptoparasites – at a heavy risk of mortality to themselves: 29% to 56% of parasite-soldier encounters ended with a live parasite and a dead soldier (Perry et al., 2004). Thus the defense behavior is altruistic towards the reproductive gall mates.

All these species of thrips are haplo-diploid, like the Hymenoptera, so that Hamilton's model for the evolution of kin selection does apply in this case. The relatedness of gall inhabitants is further enhanced by close inbreeding within the gall (Crespi, 1992).

Aphids

Species of aphids with sterile soldiers are a minority among galling aphids. The galling aphids with a soldier caste are referred to as proto-social, because all members of a colony are genetically identical (Stern and Foster, 1996).

Each gall is induced by a single fundatrix, which reproduces parthenogenetically (reviews in Wool, 2004, 2005). Aoki (1976) first called attention to the presence of two morphs of 1st-instar larvae in the galls – one of which had a heavily sclerotized body and strong forelegs (or frontal horns). These 1st-instars did not develop into reproductives (were sterile) and were observed to attack predators when artificially introduced into the gall. Soldier morphs were later discovered in a small number of galling aphids in Europe (Pemphigidae) (e.g., *Pemphigus spyrothecae*; Benton and Foster, 1992). Soldiers were thought to guard the gall entrance against clone mixing by migrants from other galls, but there is no evidence that they do (Stern and Foster, 1996).

Interestingly, the soldiers perform other duties in the gall – in particular, they remove wax-coated honeydew droplets and exuviae from the gall. Sometimes several of them cooperate to do this (Aoki, 1980; Aoki and Kurosu, 1989). Some of them lose their lives when they fall out of the gall with the honeydew. Gall cleaning is essential for the welfare of the colony but risky for the soldier. Thus this is a truly altruistic behavior. Gall cleaning may have been a precursor gall defense (Benton and Foster, 1992; Foster and Northcott, 1994).

Recently, the soldiers were observed performing another critical, altruistic chore in the colony: they repaired holes bored in the gall wall (by the experimenters, presumably also by predators). Many soldiers were seen and filmed clustering at the damaged site and sacrificing their lives in the process of sealing the hole (Kurosu et al., 2003; Pike et al., 2004). The high relatedness among individual aphids in the gall – close to 1.0 – is in line with Hamilton's kin selection theory.

Literature Cited

Alicchio, R. and L.D. Palenzona. 1974. Phenotypic variability and divergence in disruptive selection. Theoretical and Applied Genetics 45: 122-125.

Aoki, S. 1976. Occurrence of dimorphism in the first-instar larva of *Colophina clematis* (Homoptera: Aphidoidea). Kontyu 44: 130-137.

Aoki, S. 1980. Occurrence of a simple labor in a gall aphid, *Pemphigus dorocola* (Homoptera: Pemphigidae). Kontyu 48: 71-73.

Aoki, S. and U. Kurosu. 1989. Soldiers of *Astegopteryx styraci* (Homoptera: Aphidoidea) clean their gall. Japanese Journal of Entomology 57: 407-411.

Benton, T.G. and W.A. Foster. 1992. Altruistic housekeeping by a social aphid. Proceedings of the Royal Society, London, B 247: 199-202.

Craig, D.M. 1982. Group selection versus individual selection: an experimental analysis. Evolution 36: 271-282.

Crespi, B.J., 1992. Eusociality in Australian gall thrips. Nature 359: 724-726.

Darwin, C. 1874. The Descent of Man. 2nd ed., Hurst & Co., New York, USA.

Foster, W.A. and P.A. Northcott. 1994. Galls and the evolution of social berhavior in aphids. pp. 161-182. In: M.A.J. Williams (ed.), Plant Galls, Clarendon Press, Oxford, UK.

Franklin, I. and R.C. Lewontin. 1970. Is the gene the unit of selection? Genetics 65: 707-734.

Hamilton, W.D. 1964. The genetical evolution of social behavior. Journal of Theoretical Biology 7: 1-52.

Kranz, B.D., M.P. Schwartz, L.A. Mound and B.J. Crespi. 1999. Social biology and sex ratios of the eusocial gall-inducing thrips, *Kladothrips hamiltoni*. Ecological Entomology 24: 432-442.

Kranz, B.D., M.P. Schwartz, T.E. Wills, T.W. Chapman, D.C. Morris and B.J. Crespi. 2001. A fully reproductive fighting morph in a soldier clade of gall-inducing thrips (*Oncothrips morrisi*). Behavioral Ecology and Sociobiology, (on line) 10, May 2001.

Kurosu, U., S. Aoki and T. Fukatsu. 2003. Self-sacrificing gall repair by aphid nymphs. Proceedings of the Royal Society, London B 270: s12-s13.

Lester, L.J. and R.K. Selander. 1981. Genetic relatedness and the social organization of *Polistes* colonies. American Naturalist 117: 147-166.

Lewontin, R.C. 1970. The units of selection. Annual Review of Ecology and Systematics 1: 1-19.

Lewontin, R.C. 1974. The Genetic Basis of Evolutionary Change. Columbia University Press, New York, USA.

Malo, A.F., E.R.S. Roldan, J. Garde, A.J. Soler and M. Gomendio. 2005. Antlers honestly advertise sperm production and quality. Proceedings of the Royal Society, London B 272: 149-157.

Perry, S.P., T.W. Chapman, M.P. Schwartz and B.J. Crespi. 2004. Proclivity and effectiveness in gall defense by soldiers in five species of gall-inducing thrips: benefits of morphological caste dimorphism in two species (*Kladothrips intermedius* and *K. habrus*). Behavioral Ecology and Sociobiology 56: 602-610.

Pike, N. and W.A. Foster. 2004. Fortress repair in the social aphid species *Pemphigus spyrothecae*. Animal Behavior 67: 909-914.

Ritte, U. and B. Lavie. 1972. The genetic basis of dispersal behavior in the flour beetle, *Tribolium castaneum*. Canadian Journal of Genetics and Cytology 19: 717-722.

Roush, R.T. and J.A. Mackenzie. 1987. Ecological genetics of insecticide and acaricide resistance. Annual Review of Entomology 32: 361-380.

Stam, P. 1975. Linkage disequilibrium causing selection at a neutral locus in pooled *Tribolium* populations. Heredity 34: 29-38.

Stern, D.L. and W.A. Foster. 1996. The evolution of sociality in aphids: a clone's eye view. pp. 150-165. In: J. Echoe and B.J. Crespi (eds.), Social Behavior in Insects and Arachnids, Cambridge University Press, Cambridge, UK.

Van Valen, L. 1971. Group selection and the evolution of dispersal. Evolution 25: 591-598.

Williams, G.S. 1966. Adaptation and Natural Selection: a critique of some current evolutionary thought. Princeton University Press, USA.

Wade, M.J. 1977. An experimental study of group selection. Evolution 31: 134-153.

Wade, M.J. 1978. A critical review of the models of group selection. Quarterly Review of Biology 53: 101-114.

Wade, M.J. 1984. The population biology of flour beetles, *Tribolium castaneum*, after interdemic selection for increased and decreased population growth rate. Researches on Population Ecology 26: 401-415.

Wade, M.J. and C.J. Goodnight. 1991. Wright's shifting balance theory: an experimental study. Science 253: 1015-1018.

Wool, D. 1982. Family selection for DDT resistance in flour beetles (*Tribolium*: Tenebrionidae) as a model for group selection of fitness traits. Israel Journal of Entomology 16: 73-85.

Wool, D. 2004. Galling aphids: specialization, biological complexity, and variation. Annual Review of Entomology 49: 175-192.

Wool, D. 2005. Gall-inducing aphids: biology, ecology, and evolution. pp. 74-132. In: A. Raman, C.W. Schaefer, T.M. Withers (eds.), Biology, Ecology and Evolution of Gall-inducing Arthropods. Science Publishers, Enfield, USA.

Wynne-Edwards, V.G. 1962. Animal Dispersion in Relation to Social Behavior. Oliver & Boyd. Edinburgh, U.K.

Zahavi, A. 1975. Mate selection: selection for a handicap. Journal of Theoretical Biology 53: 205-214.

Zahavi, A. and A. Zahavi. 1996. (in Hebrew). Peacocks, Altruism, and the Handicap Principle. Israel Society for Protection of Nature. Tel Aviv, Israel.

CHAPTER 16

Evolution in Asexually-reproducing Populations

I inferred that uninterrupted parthenogenetic reproduction would prevent the adaptation of the species to new conditions of life. All species with purely parthenogenetic reproduction are sure to die out – not, indeed, because of any failure in meeting the existing conditions of life, but because they are incapable of ... adapting themselves to any new conditions (August Weismann, Amphimixis, 1891, p. 298).

Most species of familiar plants and animals reproduce sexually. But, despite Weismann's pessimistic prognosis, various forms of asexual reproduction are far from being rare. Reproduction by bulbs, cuttings and stolons is common in plants. In animals, cases of asexual propagation are known in invertebrates, for example in corals and sea anemones. **Parthenogenesis** is a common way of asexual reproduction in insects.

ADVANTAGES OF SEXUAL REPRODUCTION

The advantages of sexual reproduction (from a population genetics and evolutionary point of view), were realized in the late 19th century by the German zoologist, August Weismann (1834-1914). Weismann understood that the role of meiosis in cell division was to ensure that the hereditary material (the structure of which was unknown to science at the time) should not double at fertilization and accumulate in the cells: to remain the same in each individual, the amount had to be halved at the production of the gametes! Weismann suggested that the role of sexual reproduction was to create variation in the population as raw material for natural selection, by mixing of chromosomes from two parents during meiosis and fertilization (Weismann, 1891).

Many years later, with the benefit of the science of Genetics, the question of why sexual reproduction is so prevalent was discussed at length by Williams (1975). He suggested three advantageous results of sexual reproduction in evolution. (1) By mixing two different genomes at fertilization, recombination of the material during gamete formation increases variation (as suggested by Weismann). (2) When environmental conditions change – as they often do in nature – sexual reproduction increases the probability that new genotypes will be present that can withstand the new conditions. (3) Sexual reproduction increases the probability that favorable mutations will accumulate in one individual and result in a new phenotype. If, for example, three mutations are needed in order to create a favorable phenotype, the probability of all three occurring in one gamete is μ^3, which is very small – but with sexual reproduction and recombination, the three mutations may happen in different gametes and individuals – with a much higher probability (3μ) – and accumulate by recombination (Fig. 16.1).

Asexual reproduction, however, also has its advantages. The reproductive potential of a sexual population is half that of an asexual one: only half the individuals, the females, actually produce offspring. If the population reproduces sexually, the arrival of two adult individuals of opposite sex is required in order to colonize a new habitat or site, but a single asexual individual can start a colony.

When environmental conditions remain stable for a long time, it is possible that a single genotype will be favored by selection. The advantage of that genotype will be best preserved by asexual reproduction. Man has

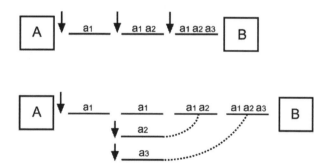

Fig. 16.1 Two pathways for accumulation of three favorable mutations to create a favorable phenotype. Top: the mutations occur in sequence in the same individual. Bottom: mutations in different individuals accumulate by recombination.

long realized that favorable varieties of fruit trees are best propagated by asexual means, like cuttings and grafting, rather than from seeds. However, if the environment changes drastically, the limited variation by asexual reproduction may lead to extinction.

An asexual population may suffer great mortality without a change in genotype frequencies. Many aphids (genus *Uroleucon*) feeding on lettuce (Compositae) die because contact of the legs or proboscis of the aphid with the stem, flower stalks or buds causes exudation of latex, which traps the insect (Dussourd, 1995). In experiments, 80-90% of the feeding aphids were thus trapped and most of them died, although trapped young aphids were able to molt and walk away, leaving the old skin behind. Dussourd suggests that feeding on lettuce may be advantageous for the asexual aphids, since many of their parasites and predators – which reproduce sexually – were also trapped!

PARTHENOGENESIS

Parthenogenesis is a reproductive system in which mothers produce daughters without fertilization. Parthenogenesis was reported in some invertebrate – rotifers, small crustaceans (e.g., *Daphnia*) and some insects, notably aphids (Dixon, 1998). That aphids reproduce without fertilization was discovered by Charles Bonnet in the middle of the18th century, and the anatomical and embryological details of their parthenogenetic reproduction were described by T.H. Huxley (1858). Parthenogenesis is rare in vertebrates, but was reported in some fishes, amphibians and reptiles (review in the American Zoologist, 11 (1971)).

There are several cytologically-different mechanisms of parthenogenesis in insects (Suomalainen, 1950, 1962). The most extreme is called apomictic parthenogenesis. No meiosis occurs in egg cell formation (therefore no recombination is possible) and diploid daughters are produced without fertilization. All daughters are consequently genetically identical and have their mother's genotype (except the rare occurrence of somatic mutations). This is the prevalent reproduction method in parthenogenetic aphids and in some beetles.

In automictic parthenogenesis there is no fertilization, but meiosis does occur during egg development, and recombination is possible between chromosomes. The haploid egg cell unites with one of the polar bodies to create a new diploid individual.

A third parthenogenetic mechanism, which results in unisexual populations of inter-specific hybrids, was discovered in fishes *(Poeciliopsis)* in central America (Vrijenhoek et al., 1977).

Cyclic Parthenogenesis

In *Daphnia* and particularly in aphids, a sequence of a few parthenogenetic generations is interrupted by a single sexual phase. This often occurs when or just before the environment changes for the worse, as when temporary rain pools dry up (in *Daphnia*), or before leaf senescence and the onset of winter in European aphids. The sexual generation results in the formation of durable, aestivating or over-wintering eggs, which may pass the stress period unharmed.

For example, during most of the summer, gall-inducing aphids (Pemphigidae: Fordinae) in the Mediterranean area reproduce 3-4 parthenogenetic generations in galls on their host trees, *Pistacia* (Anacardiaceae). With the approach of the autumn leaf fall, the aphids escape from the galls and reproduce parthenogenetically during the winter on the roots of their secondary hosts, grasses. But when the environment changes again – in spring – their offspring (sexuparae) fly to the *Pistacia* trees and give rise to males and females, which mate and produce fertilized eggs. The eggs remain dormant for the duration of the hot and dry summer. A nymph hatches in the following spring and induces a new gall (reviews in Wool, 2004, 2005).

Electrophoretic variation in *Daphnia* populations in England was investigated in two habitats: continuous, permanent bodies of water, and temporary winter rain-water ponds. In the permanent habitats, reproduction was parthenogenetic with periodic bursts of sexual reproduction. In the temporary ponds, there was a cyclical alternation between the two reproductive strategies: a change to sexual reproduction occurred and males appeared when conditions became worse. After mating, each female produced two permanent eggs, which sank to the bottom mud, and could withstand desiccation when the pond dried up.

Cyclic parthenogenesis seems to be the ideal reproductive method – combining the advantages of both sexual and asexual reproduction. It is remarkable that it is less widespread than might be expected. In Canada, most of the 77 populations of *Daphnia* investigated were obligatorily parthenogenetic. Only a few were cyclically-parthenogenetic (Hebert et al., 1989; Crease et al., 1989).

PARTHENOGENESIS, POLYPLOIDY AND VARIATION

Contrary to the opinion that asexual reproduction may be a dead-end in evolution because it limits variation, a remarkable fact about parthenogenesis is that it protects variation. Any mutation which occurs in a parthenogenetic, diploid individual is protected in the heterozygous state.

Even deleterious recessive mutations can accumulate, since in the absence of recombination they cannot be exposed to selection in a homozygote.

Proof of the above statement is provided by molecular sequence comparisons of parthenogenetic and sexual lineages of aphids. These studies confirm that asexual lineages show excess of heterozygosity compared with sexual lineages. For example, in a wide-scale study of the aphid *Rhopalosiphum padi* in France, the sexual lineages showed greater electrophoretic allele diversity – twice as many alleles and genotypes – as the asexual lineages, but at the molecular level – using DNA microsatellite comparisons – the asexual lineages showed high heterozygosity at most loci (Delmotte et al., 2002).

Major chromosomal changes are preserved in some parthenogenetic aphid populations and give rise to viable strains. An important chromosomal phenomenon – quite common in plants – is the multiplication of the numbers of chromosomes per cell, resulting in triploid, tetraploid or even higher-ploid individuals. Polyploidy results when the chromosomes in mitosis do not migrate into separate cells (this can be induced experimentally by applying appropriate chemicals to dividing cells). Polyploidy usually creates a reproductive barrier between the polyploid and its diploid parent species due to unbalanced genetic material in hybrid gametes. Polyploids can, however, be propagated asexually.

Polyploidy is common in cultivated plants like wheat and barley. It is far less common in animals. Nevertheless, of 47 parthenogenetic strains and species of insects listed by Suomalainen (1962), a large majority were polyploids. Of 42 strains or species of curculionid beetles, 25 were triploid, 12 were tetraploid, and four pentaploid; only one species was diploid. It is not clear what came first: the change in ploidy number- or parthenogenetic reproduction (Suomalainen et al., 1976).

It is remarkable that in most cases the parthenogenetic strains had a much wider geographic range than closely-related diploid species, from which they had probably evolved. The polyploids appeared able to utilize a wider range of habitats and climates. One explanation for this may be historical: Suomalainen (1950, 1962) suggested that during the ice age, the available habitats for the insects were limited to a few ice-free refuges. As the ice retreated and new habitats became available, the reproductive advantage of parthenogenetic polyploids enabled them to colonize wider areas faster than the original sexual diploids. This explanation is supported by data on several species of beetles and moths.

A similar phenomenon was reported in plants. In Colorado, the obligatorily-asexual species of *Antennaria parviflora* (Asteraceae) were found up to 3500 m altitude, while the sexually-reproducing plants rarely grew above 2750 m (Bierzychudek, 1989). In detailed laboratory experiments, subunits derived from the same individual plants were subjected to six different treatments of irrigation and temperature. The asexual plants produced more biomass, more flowers, and survived better than the sexual plants.

EVOLUTION OF PARTHENOGENESIS AND POLYPLOIDY

The evolutionary importance of parthenogenesis is that it permits instantaneous speciation... However, it generally gains only short-term advantage...

Though superficially appearing a "more primitive" type of reproduction, parthenogenesis in recent animals is evidently in all cases secondarily derived from sexual reproduction (Mayr, 1963, p. 411).

Most investigators agree that parthenogenetic species evolved from sexual progenitors. Parthenogenetic species of animals are not necessarily related taxonomically, suggesting multiple origins of this reproductive system (Suomalainen, 1962; Parker et al., 1977).

Parthenogenesis in insects involves two steps: 1) oviposition by the females without previous fertilization of the eggs, and 2) the unfertilized eggs continue to develop into larvae and adults.

In haplo-diploid species, unfertilized eggs develop into males. Oviposition without fertilization is known in a number of species of *Drosophila* and in flour beetles (*Tribolium*), but as a rule unfertilized eggs do not develop into viable embryos. In some species, however, eggs occasionally do develop spontaneously. If this tendency has a genetic basis, the ability to hatch from unfertilized eggs may have an advantage, e.g., following a temporary, accidental loss of all males in a small isolated population, or when a single unfertilized female arrives at a new site: if some of her eggs develop into adults, she may start a colony. Flour beetles were used as models to understand the genetic control of the phenomenon (Lavie et al., 1978; Orozco and Bell, 1974).

In *Drosophila*, about 30 species are known where embryos begin to develop in a small proportion of unfertilized eggs. In five of these species, 1 - 5% of these eggs developed to adults. In one exceptional species – *D. mangabieri* – about 60% of the unfertilized eggs began development spontaneously, and many of them eventually became adults. Only females are known in this species. Females collected in nature produced more than 4000 offspring in the laboratory, all female (Carson, 1960).

Carson (1967) selected for an increase in proportion of parthenogenetic eggs in *D. mercatorum* in the laboratory. The proportion of unfertilized eggs developing parthenogenetically increased from 1% to 6.4% – not a dramatic increase but still evidence that the trait had a genetic basis in that species. Parthenogenesis in *D. mercatorum* is automictic (with meiosis).

GENETIC VARIATION IN PARTHENOGENETIC POPULATIONS

Parthenogenesis was for a long time considered a dead end in evolution: the absence of recombination is expected to reduce genetic variation and restrict the operation of natural selection. On the other hand, mutations may accumulate in parthenogenetic populations, and they may not be genetically impoverished after all.

One of the first examples of genetic variation in parthenogenetic species was reported by Carson (1960). He examined the salivary gland chromosomes of 270 larvae of the parthenogenetic species *D. mangabieri*, mentioned above, and found that all of them were heterozygous for three chromosomal inversions. Suomalainen et al. (1976) reported that electrophoretic variation (at 12-15 isozyme loci) in parthenogenetic species and strains of curculionid beetles and in a psychid moth, was no smaller than in their sexual counterparts. An exception was a chrysomelid beetle, *Adoxus*, where most loci were monomorphic in all populations and strains. The beetles are good fliers and can disperse over large distances (in contrast to the flightless curculionids, whose flightlessness facilitated differentiation).

Electrophoretic variation in parthenogenetic aphids is very limited (Wool et al., 1978). Very little variation in electrophoretic mobility of esterase isozymes was discovered in 64 lines of *Myzus persicae* collected from many sources in England. On the other hand, there was considerable variation in the intensity of activity of these isozymes. Activity variation was related to organophosphorous insecticide resistance of the source populations. The assumption that a larger quantity of the enzyme was produced in the resistant strains due to amplification of the gene coding for the enzyme (Bunting and van Emden, 1980), was later confirmed by Devonshire and Field (1991). The greater number of enzyme molecules sequester more of the poison molecules, enabling the individual aphids with the high-activity esterase to survive (Field et al., 1988).

The limited electrophoretic mobility variation in *M. persicae* – confirmed later in other aphid species as well – may be due to the combination of a fast reproductive rate and the flight ability of the winged stages in the life cycle of aphids. A few females landing in a large field can rapidly establish genetically-homogeneous colonies which will spread over a large area.

A case in point is the rapid spread of the parthenogenetic aphid, *Elatobium abietinum*, in New Zealand (Nicol et al., 1998). This aphid has caused mortality of sitka spruce trees in large commercial plantations in England since the 19th century. Small forested areas of sitka spruce were planted by European settlers in New Zealand, and aphids were probably transferred inadvertently with the trees. Nicol et al. (1998) used molecular markers to study genetic variation in *E. abietinum* samples from New Zealand and England. In England, where the maximal distance between sites was 240 km, a sample of 40 aphids contained 28 different genotypes. In a similar sample from New Zealand, with 1200km between sites, there was no variation at all – all the sampled aphids were genetically identical, despite the use of many primers. The origin of the New Zealand aphids is undoubtedly European, possibly even from a single immigrant, since reproduction in the New Zealand aphid is strictly parthenogenetic. This mating system is very favorable for colonization if an immigrant can find a site with plenty of food and no competitors, as was apparently the case for *E. abietinum* in New Zealand.

MORPHOLOGICAL VARIATION IN PARTHENOGENETIC SPECIES

Most measurable morphological characters are affected by both genetic and environmental factors. With appropriate statistical techniques it is possible to estimate the relative contribution of the two sources of variation to individual characters. Such analyses were carried out in at least two groups of parthenogenetic insects: gall-forming aphids and grasshoppers.

The gall-forming aphids are cyclically parthenogenetic and have complex life cycles (review in Wool, 2004, 2005). The fundatrix nymph induces a gall in which she reproduces. Since parthrenogenesis in aphids is apomictic, all her daughters in the gall are genetically identical. Morphological variation among sisters in the gall must be entirely due to developmental and environmental factors during aphid ontogeny. The differences among the means of the characters in different galls on the same tree or branch can have a genetic component, because each gall is induced by a different fundatrix. Using a 'nested' (hierarchical) statistical analysis of variance (Sokal and Rohlf, 1995), the relative magnitude of the genetic and environmental components of variation were estimated for three species of galling aphids on *Pistacia* (Wool, 1977). Table 16.1 lists the results for one species (*Geoica wertheimae*) sampled in 215 galls from 35 sites. The proportion of non-genetic and genetic variation of different characters is not the same. The genetic component (among galls) appears to be rather similar, but the within-gall environmental component varies between 6% and 30% in different characters.

Table 16.1 Variance components of morphological characters of the gall-inducing aphid, *Geoica wertheimae*. Character codes (in parenthesis) are given as a connection to the published papers. The components of variation among sites, and among trees within sites, are omitted (this is why the values in each line do not add up to 100%). Data from Wool, 1977

Character	% variance	
	Among galls	within galls
Wing length (WL)	41.5	8.6
Wing width (WW)	23.5	30.4
Head width (HW)	34.6	23.8
Thorax width (TW)	37.8	21.0
3rd antennal segment (A3)	54.8	10.4
4th antennal segment (A4)	45.8	20.1
5th antennal segment (A5)	47.7	20.5
6th antennal segment (A6)	16.0	27.5
Sensoria on A3 (s1)	42.5	13.7
Sensorial on A4 (s2)	33.5	27.3
Femur, foreleg (F1)	35.2	8.6
Tibia, foreleg (Ti1)	32.7	7.3
Tarsus, foreleg (Tar1)	32.4	18.6
Femur, midleg (F2)	35.6	9.0
Tibia, midleg (Ti2)	32.8	6.0
Tarsus, midleg (Tar2)	33.6	14.5
Femur, hindleg (F3)	37.5	8.2
Tibia, hindleg (Ti3)	34.2	8.0
Tarsus, hindleg (Tar3)	34.2	10.9

Parthenogenetic grasshoppers (genus *Warramaba*) are very common in some areas of Australia (Atchley, 1977a, b). Four species of the genus are sexual, but there is a parthenogenetic strain *W. virgo* – of which only females are known. This is a flightless grasshopper, which feeds on the leaves of different species of wattle *(Acacia)*. Cytologically, parthenogenesis in this species is automictic – with meiosis and remerging of the nuclei in egg formation – but no crossing-over was detected, and the author considers that no recombination takes place.

Variation in 13 morphological characters showed that 70% of the variance was within clones (non-genetic) and only 30% was among clone means. The partitioning of variation was similar in the sexual and asexual populations of the grasshopper.

Literature Cited

Atchley, W.R. 1977a. Biological variability in the parthenogenetic grasshopper *Warramaba virgo* (Key) and its sexual relatives. I. The eastern-Australian populations. Evolution 31: 782-799.

Atchley, W.R. 1977b. Evolutionary consequences of parthenogenesis: evidence from the *Warramaba virgo* complex. Proceedings of the National Academy of Science, USA 74: 1130-1134.

Bierzychudek, P. 1989. Environmental sensitivity of sexual and apomictic *Antennaria*: do apomicts have general-purpose genotypes? Evolution 43: 1456-1466.

Bunting, S. and H.F. Van Emden, 1980. Rapid response to selection for increased esterase activity of small populations of an apomictic clone of *Myzus persicae*. Nature 285: 502-503.

Carson, H.L. 1960. Fixed heterozygosity in a parthenogenetic species of *Drosophila*. University of Texas publications (1960), 2: 55-62.

Carson, H.L. 1967. Selection for parthenogenesis in *Drosophila mercatorum*. Genetics, 55: 157-171.

Crease, T.J., D.J. Stanton and P.D.N. Hebert. 1989. Polyphyletic origins of asexuality in *Daphnia pulex*. II. Mitochondrial DNA variation. Evolution 43: 1016-1026.

Delmotte, F., N. Leterme, J-P. Gauthier, C. Rispe and J-C. Simon. 2002. Genetic architecture of sexual and asexual populations of the aphid *Rhopalosiphum padi*, based on allozymes and microsatellite loci. Molecular Ecology 11: 711-723.

Devonshire, A.L. and L.M. Field. 1991. Gene amplification and insecticide resistance. Annual Review of Entomology 36: 1-23.

Dixon, A.F.G. 1998. Aphid Ecology – an optimization approach. Prentice-Hall, London, UK.

Dussourd, D.E. 1995. Entrapment of aphids and whiteflies in lettuce latex. Annals of the Entomological Society of America 88: 163-172.

Field, L.M., A.L. Devonshire and B.G. Forde. 1988. Molecular evidence that insecticide resistance in peach-potato aphid (*Myzus persicae* Sulzer) results from amplification of an esterase gene. Biochemical Journal 251: 309-312.

Hebert, P.D.N., M.J. Beaton, S.S. Schwartz and D.J. Stanton. 1989. Polyphyletic origins of asexuality in *Daphnia pulex*. I. Breeding system variation and levels of clonal diversity. Evolution 43: 1004-1015.

Huxley, T.H. 1858. On the agamic reproduction of *Aphis*. Transactions of the Linnean Society, Part I, 22: 193-220; Part II, 22: 221-241.

Lavie, B., U. Ritte and R. Moav. 1978. The genetic basis of egg lay response to conditioned medium in the flour beetle *Tribolium castaneum*. I. Two-way selection. Theoretical and Applied Genetics 52: 193-199.

Mayr, E. 1963. Animal Species and Evolution. Harvard University Press, Boston, USA.

Nicol, D., K.F. Armstrong, S.D. Wratten, P.J. Walsh, N.A. Straw, C.M. Cameron, C. Lahman and C.M. Framton. 1998. Genetic diversity in an introduced pest, the green spruce aphid *Elatobium abietinum* (Hemiptera: Aphididae) in New Zealand and the United Kingdom. Bulletin of Entomological Reaearch 88: 537-543.

Orozco, F. and A.E. Bell. 1974. A genetic analysis of egg laying of *Tribolium* in optimal and stress environments. Canadian Journal of Genetics and Cytology 16: 49-60.

Parker, E.D., R.K. Selander, R.O. Hudson and L.J. Lester. 1977. Genetic diversity in colonizing parthenogenetic cockroaches. Evolution 31: 836-842.

Sokal, R.R. and F.J. Rohlf. 1995. Biometry. 3rd ed. Freeman, New York, USA.

Suomalainen, E. 1950. Parthenogenesis in animals. Advances in Genetics 3: 193-253.

Suomalainen, E. 1962. Significance of parthenogenesis in the evolution of insects. Annual Review of Entomology 7: 349-366.

Suomalainen, E., A. Saura and J. Lokki. 1976. Evolution in parthenogenetic insects. Evolutionary Biology 9: 209-257.

Vrijenhoek, R.C., R.A. Angus and R.J. Schultz. 1977. Variation and heterozygosity in sexually versus clonally reproducing populations of *Poecilliopsis*. Evolution 31: 767-781.

Weismann, A. 1891. The significance of sexual reproduction. Essays on Heredity, 2nd ed. Vol. I. Clarendon Press, Oxford, UK.

Weismann, A. 1891. Amphimixis. Essays on Heredity, 2nd ed. Vol. II. Clarendon Press, Oxford, UK.

Williams, G.C. 1975. Sex and Evolution. Princeton University Press, Philadelphia, USA.

Wool, D. 1977. Genetic and environmental components of morphologuical variation in gall-forming aphids (Homoptera, Aphididae, Fordinae) in relation to climate. Journal of Animal Ecology 46: 875-889.

Wool, D., S. Bunting and H.F. Van Emden. 1978. Electrophoretic study of British *Myzus persicae* (Sulz.) (Hemiptera, Aphididae). Biochemical Genetics 16: 987-1006.

Wool, D. 2004. Galling aphids: specialization, biological complexity, and variation. Annual Review of Entomology 49: 175-192.

Wool, D. 2005. Gall-inducing aphids: biology, ecology, and evolution. pp. 73-132. In: A. Raman, C.W. Schaefer, and T.M. Withers (eds.), Biology, Ecology and Evolution of Gall-inducing Arthropods. Science Publishers, Enfield, USA.

CHAPTER 17

Laboratory Populations as Models for Natural Selection

ADVANTAGES AND LIMITATIONS

Evolutionary processes result from interactions of genotypes with environmental factors – physical, like temperature, rainfall and soil variables – and biological, like predators, parasites and competition with other genotypes. It is difficult to study such interactions in nature. This is why the number of documented cases of evolutionary change in natural populations (Chapter 12) is so small. To estimate the relative importance of different interactions that might be encountered in nature, controllable laboratory populations are used as models.

Ecological-genetic models examine, for example, whether different genotypes require different environmental resources; whether different genotypes respond differently to environmental physical conditions; whether genotypes prefer different types of food or habitat; and whether these characteristics change with time.

Strains which carry visible, easily recognizable markers are used in laboratory models, and it is possible to follow changes in allele frequencies in response to specified environments. The choice of organisms with a relatively short generation time enables the observation of evolutionary processes in progress within reasonable experimental periods. The understanding of such interactions in the laboratory may provide the key to the interpretation of natural processes that occurred in the past and only the end results of which are observable – and the prediction, within limits, of possible future trends.

It is important to remember that the number of suitable organisms for such experiments is necessarily limited. Long-term experiments are often performed with *Drosophila* and with flour beetles, *Tribolium*. Very large populations of these insects can be reared in small containers with simple food and maintenance. Interpretation of the laboratory models, however, must be made with caution because both the design and the results of the biological simulation may be affected by the biological requirements and characteristics of the model organism.

GENETIC AND EPIGENETIC INTERACTIONS WITHIN POPULATIONS

Epigenetic interactions may affect the fate of individuals.The Russian geneticist Timofeev-Resovsky produced, by appropriate crosses, laboratory mutant strains of *Drosophila* of two kinds: one group carried each mutation singly on a different chromosome. In the other group, each chromosome carried two of the same mutations. All but one of the single mutations had deleterious effects on fitness (most of them caused reduced survival). It was expected that survival of the double-mutant strains would be even lower – a product of the survival rates of the two mutations – but this was not the case: in some combinations, survival of the double mutants was even higher than that of the single mutants (Table 17.1; extracted from Wallace, 1969).

Table 17.1 Percentage survival of *Drosophila funebris* strains carrying one or two mutations. Wild-type survival was taken as 100. Expected survival of a double mutant is the product of survival of the two single mutations. Data from Timofeev-Resovsky, cited by Wallace, 1969

First mutation alone		second mutation alone		double mutant		expected
Code	survival	code	survival	code	survival	survival
Ev	104	ab	69	ev+ab	84	72
Ev	104	sn	79	ev+sn	103	82
L_3	74	bb	85	l_3+bb	85	63
Ev	104	bb	85	ev+bb	86	88
m	69	sn	79	m+sn	67	55
m	69	ab	69	m+ab	83	48
m	69	bb	85	m+bb	97	59
ab	69	l_3	74	l_3+ab	59	51
ab	69	bb	85	ab+bb	79	59

GENETIC FACILITATION

One kind of genotype-genotype interaction within populations – sometimes mediated by the environment – was termed **genetic facilitation**. Lewontin (1955) maintained *Drosophila* strains in single-strain populations, and in competition with a standard white-eyed strain. The expectation was that competition would have a negative effect because food or space (or both) could become limiting. Surprisingly, in 10 of 19 interactions, survival was higher in the mixed than in the single-strain populations: the white-eye strain seemed to have a positive effect on the survival of its competitor (in six interactions survival was reduced, and in three others there was no difference).

A similar case was reported in flour beetles. The presence of a mutant strain, sooty (**ss**), in mixed cultures, improved the growth rate and final pupal weight of the other genotypes, **+s** and **++**. The effect was more pronounced the higher the frequency of **ss** in the population (Sokal and Huber, 1963).

Genetic facilitation was investigated in detail in houseflies. Bryant (1969) used a wild type strain and a green-eyed mutant, **ge**. The mutant grew better and reached heavier weight in mixed cultures with the wild-type strain than in single-strain **ge** cultures. It was not necessary to keep the wild-type larvae alive: the **ge** strain fared better when growing in the used wild-type culture medium when wild-type larvae were killed by freezing before introducing **ge**. Further quantitative research showed that the wild-type larvae, when alive, encouraged the development of yeast in the medium – on which fly maggots of both strains feed. The **ge** in the used culture medium were thus given an enriched diet.

Another case of genetic facilitation involved the Amylase-coding gene in *Drosophila*. (Amylases are enzymes that break down starch polymers into disaccharides, mostly maltose.) A 'null' mutant *Drosophila* strain was unable to grow on standard medium when starch was the only source of carbon. But in competition with a wild-type strain, the null mutant grew well (Haj-Ahmad and Hickey, 1982). The normal larvae produced excessive amounts of amylase, some of which was excreted into the medium. The enzyme pre-digested the starch in the medium into sugars – a process which normally occurs inside the gut – and the null maggots thrived on the products.

MODELING EVOLUTIONARY PROCESSES WITH LABORATORY POPULATIONS

There are two approaches to the laboratory study of evolutionary processes. The 'synthetic' approach is to test the effects of environmental variables on biological parameters – like survival, developmental time or fecundity – on each genotype (strain) separately. Once a large body of data of this nature has accumulated, one can combine the information to simulate the effect of all factors simultaneously on a mixed population. A modern way of 'synthesizing' the combined effects of many factors is to use computer simulation. The simulation programs predict the outcome of the simultaneous activity of the known factors on the future size and /or composition of the population. The predictions can be tested later on by new experiments. This approach has the advantage that new sets of data may be added, or parameter values can be changed, and the entire array can be reanalyzed generating new testable working hypotheses and predictions.

The 'synthetic' approach has the disadvantage that collecting the data requires a heavy work load. The effect of each factor is studied separately in independent experiments, and the combined effect of different factors is not necessarily the sum of their separate effects. Further, the values of the biological parameters assigned to each genotype in the simulation programs are constant – while in reality parameters may be expected to change as the population evolves. Finally, with this approach it is not possible to estimate the importance of factors that were not expected a priori.

The second way of constructing an evolutionary model in the laboratory is the 'analytic' approach. One can set up experimental populations with known initial genetic composition, and place them in controllable environments. The changes in the frequencies of genotypes are monitored in a sequence of generations without interfering with the populations for the duration of the experiment. When the experiment is over, the temporal changes in allele frequencies are analyzed, and working hypotheses are formulated about the forces that had caused the changes. These hypotheses can be tested in suitable, short-term experiments.

The advantage of the analytic approach is that the experimenter does not disturb the processes, and the population runs its course 'naturally'. Many variables can be evaluated while interpreting the results – including some which were not anticipated at the start. Analytical experiments are thus closer to natural processes, with the further advantage that the initial conditions are known, unlike the situation in nature.

The disadvantage of the analytic approach is that the processes which occur in the population cannot be manipulated by the experimenter while the experiment is in progress. Once the main experiment is over, it cannot be repeated, and even if exactly the same initial conditions are guaranteed, the results may be quite different (Lewontin, 1966).

The examples described below illustrate the synthetic approach and the modeling of evolutionary processes in the laboratory. They also illustrate difficulties which may be encountered in interpreting the results.

THE ECOLOGY OF SELECTION IN *TRIBOLIUM* POPULATIONS

Sokal and Sonleitner (1968) carried out a long-term series of experiments with the flour beetle, *Tribolium castaneum*, lasting 35 generations (about three years). They used a wild-type strain (reddish-brown) and a black strain (homozygous at the semi-dominant **b** locus) to construct three kinds of <u>mixed</u> populations, each replicated five times. Three replicates each of the <u>pure</u> populations ++ and **bb** were run in parallel. Each replicate was held in a jar with 40 g of food medium.

Methods

Mixed population were initiated with 200 eggs in Hardy-Weinberg equilibrium frequencies, at initial black allele frequencies of 0.25, 0.5, and 0.75. Eggs were obtained from 'egg farms' of ++, **bb** and heterozygotes +**b** (from the cross of ++ with **bb**). As **b** is semi-dominant, the heterozygote +**b** can be recognized in adult beetles by its 'bronze' color.

Although the experiment was continuous, discrete generations were maintained. At the end of each generation, the adults of each replicate were sifted out of the culture medium and transferred to an oviposition jar to lay eggs for three days. These eggs were counted and started a new generation. Then the adults that laid these eggs were sorted by genotype, counted and discarded.

Replicates of 200 small, medium and large larvae, and pupae were sampled from a second cohort of eggs laid by the same parents every generation to study intra-generation processes. These were reared to the adult stage and sorted by genotype, without disrupting the processes in the main line.

Results

All populations increased rapidly in size to several thousand individuals in just 4-6 generations. After reaching this size, the mixed populations remained stable in size, as the increasing density also increased mortality and negatively affected egg input. The mixed populations were twice as large as the pure ++ and **bb** – indicating that a mixture of genotypes had an advantage in utilizing the same amount of food and space compared with single-genotype populations. The **bb** population was the smallest and most negatively affected by density (Fig. 17.1).

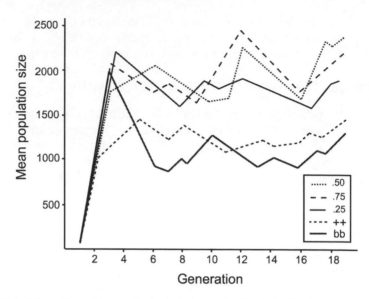

Fig. 17.1 Mean sizes of pure and mixed-strain populations in the first 19 generations of the long-term experiment. Mixed populations – marked by the initial black allele frequency – were consistently larger than the single-strain populations (after Sokal and Sonleitner, 1968).

A most interesting result, however, was a temporal change in the genetic composition in the mixed populations. Allele frequency of **b** increased from the initial q_b =0.25 to about 0.6 in 6-8 generations. In the q_b = 0.75 lines, q_b decreased – first steeply, then more slowly (Fig. 17.2). But in each generation, adult <u>genotype</u> frequencies did not depart significantly from equilibrium (note that the generation means remained very close to the parabola in the de-Finetti diagram shown in chapter 3 (Fig. 3.3).

An agreement of <u>genotype</u> frequencies with expectations means that the Hardy-Weinberg assumptions are met – but the directional changes in <u>allele</u> frequencies show that selection was favoring the **b** allele!

The reason for the apparent contradiction was discovered in the short-term within-generation experiments. Allele frequencies in the samples of the immature stages showed that very considerable genetic changes occurred within a generation – mostly due to cannibalistic predation of larvae and adults on eggs and pupae (cannibalism is a well-documented phenomenon in *Tribolium*). Selection did not favor the same genotypes at the different developmental stages, and the end result was that adult genotype frequencies at the end of a generation were not statistically different from equilibrium – while selection during the immature stages affected a slow, net directional change in allele frequency between generations.

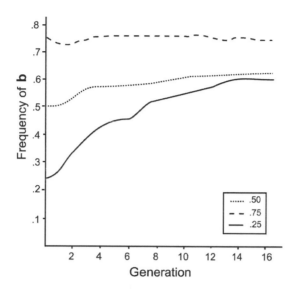

Fig. 17.2   Mean marker allele frequency (q_b) changes in the mixed lines in the first 17 generations of the long-term experiment (after Sokal and Sonleitner (1968)).

This conclusion is important. Most of the mortality in natural populations takes place in the immature stages – Darwin was well aware of the fact. In some organisms, the numbers of eggs laid are several orders of magnitude larger than the number of surviving adults. Most of the mortality of immatures is probably independent of genotype. In most cases, however, there are no genetic markers in immatures enabling observations on selective mortality before the adult stage.

The **b** locus is only one of hundreds or thousands of loci in the genome of *Tribolium*. Genotypes at this locus probably do not affect fitness (Stam, 1975). After generation 19, Sokal and Sonleitner reconstructed 'pure' **bb** and **++** strains from the mixed populations and started the experiment over again from samples of 200 eggs in Hardy-Weinberg proportions. The differences that were observed between populations in the original experiment were no longer detectable in the reconstructed populations. The allele frequency q_b did not change in 15 additional generations. The observed differences in the original experiment were due to some other sites in the genome, which were reshuffled by mating in the mixed cultures. The marker **b** in the reconstituted strains no longer represented the same ensemble of genes as in the first experiment.

The composition of the mixed populations was changing with time by selection. New genotypes may have been added by recombination.

Population size and density changed greatly with time, and may have affected the performance of the genotypes in various ways. Even cannibalism, which is rarely selective in *Tribolium*, may change allele frequencies because different genotypes pupate at different times, and hence become vulnerable to differential cannibalism by adults (Wool, 1976).

The changes in frequency of the **b** allele in Sokal and Sonleitner's experiments was later shown to be due, in part, to non-random mating success of the genotypes. An advantage of +**b** heterozygous males was discovered in mate-choice experiments. Females of either ++ or **bb** were paired with males of two genotypes at a time (+**b**, **bb** or +**b**, ++), for a short period (< 2 h, to minimize the chance of repeated mating). Each female was held individually in a vial with flour and the eggs laid were allowed to develop to adults (Wool, 1970). There were significant deviations from equilibrium in the offspring genotypes. When plotted in the hexagonal diagrams (Chapter 3) these deviations showed a definite pattern: +**b** males had a mating advantage over the homozygous male genotypes. For example, when the males were +**b** and ++, 18 of 20 replicates showed an excess of +**b** and **bb** offspring of ++ females (Fig. 17.3) – when the only source of the **b** allele were the +**b** males. When the males were +**b** and **bb**, all 20 replicates showed an excess of ++ offspring – the source of + were

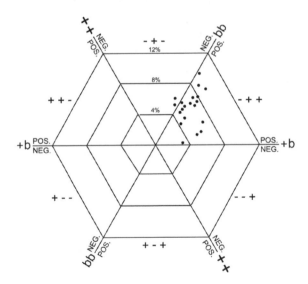

Fig.17.3 A non-random deviation pattern in 20 replicate samples of offspring of mating of +**b** and ++ males to ++ females. In 18 cases there was indication of mating advantage of +**b** males. A similar advantage was observed when +**b** and **bb** competed for ++ females: all 20 replicates showed an excess of ++ (after Wool, 1970).

again the **+b** males. This advantage was confirmed by direct observations of mating behavior (Wool, 1970).

BOTTLENECKS, INBREEDING AND FITNESS

Sudden, deleterious environmental changes may drastically reduce population size. A drastic reduction in population size, from which the population eventually recovers, is called a **bottleneck**.

Bottlenecks have genetic as well as ecological consequences: a population which recovers from a bottleneck will by definition be small, and therefore inbred. It may eventually reproduce and become very large, but will necessarily contain a more limited gene pool than before the bottleneck.

Repeated bottlenecks were imposed on populations of flour beetles in two environments: one constant and near-optimal, the other variable, unpredictable, and deleterious. Each of ten source populations was founded by a single pair of adults, thereby creating a severe initial bottleneck (Fig. 17.4).

From each source, 10 experimental lines were propagated from eggs laid in one week by a single fertile pair of siblings from the preceding

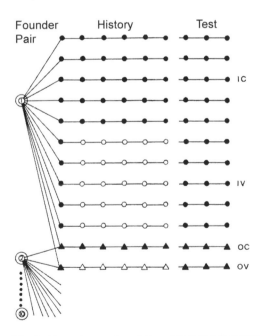

Fig. 17.4 Design of the bottleneck experiment (from Wool and Sverdlov, 1976). For details see text.

generation – again imposing a bottleneck. Five replicates were held in the constant environment (IC) for the duration of the experiment, and five were given randomly-selected deleterious treatments – simulating catastrophes – during the larval period (IV). Treatments were applied during the first six generations. In the last three generations, IV lines were transferred to the constant conditions for comparison with IC (Fig. 17.4). Two other lines, OC and OV – with five replicates each – were propagated by crosses between lines as a control for the effects of inbreeding (Wool and Sverdlov, 1976). Fitness was estimated as the number of adults per replicate at the end of a generation.

Bottlenecked experimental lines produced smaller numbers of offspring than the outbred controls. The number of infertile pairs in IC and IV increased with time after the bottleneck, as expected from inbreeding depression. But when the surviving lines in the last three generations (held in the constant environment) – were grouped by the source populations, the replicates that came from IV were on average better fit than their sisters which received no deleterious treatments (Fig. 17.5).

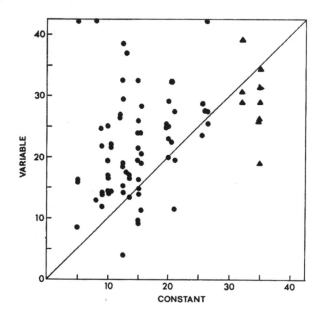

Fig. 17.5 Comparative fitness of IC and IV surviving lines. Each filled circle is the bivariate mean of replicates from the same source, held in two environments in the first six generations of the experiment. Triangles are outbred lines. Abscissa: mean number of offspring in surviving lines of IC. Ordinate: mean number of offspring in surviving lines of IV. The majority of the circles are above the diagonal (equality) line: IV lines surviving treatments had higher fitness than corresponding IC lines (from Wool and Sverdlov, 1976).

An unplanned demonstration of the superior fitness of the surviving IV populations was obtained when the experiment was interrupted by war. Samples of 20 adults from each replicate were put in 20 g of medium just before the interruption, but the cultures were neglected for almost a year. When examined, the 10 jars with IV beetles contained little, heavily contaminated medium. Census revealed a mean of 600 adults per jar, all

Fig. 17.6 Representative surviving populations after a year of neglect. The jar on the right in each photograph is from IC. The others are from IV.

dead. The four jars with the IC beetles were about half-full of medium in relatively good condition, and contained a mean of 200 adults per jar – some of them alive, and even some live larvae (Fig. 17.6).

The IV populations that survived the variable-environment 'catastrophes' were apparently selected for higher reproductive rate (r-selection). They increased quickly in size, consumed all the medium – and died of starvation because the food was not renewed. The IC lines reproduced slowly, produced fewer adults and still had some medium to feed on when re-examined after 11 months.

(Take-home message: when the food supply is limited or not renewed quickly enough, a low reproductive rate may be an advantage – while a higher reproductive rate can lead to an early extinction!)

A MODEL FOR THE DIFFERENTIATION OF ISLAND POPULATIONS

Imagine a group of islands (an archipelago), recently colonized by migrants from a distant, polymorphic population on a mainland. Founders arriving at different islands may be few and genetically different. If inter-island migration is limited, each island population may follow a different evolutionary course, leading to adaptive radiation and speciation.

Genetic differences among island populations were attributed to founder effects and selection (e.g., Ayala et al., 1971; Gorman et al., 1975). How important are founder effects, inbreeding and selection in the evolution of island populations? An experiment was set up to model this scenario, with flour beetles, *Tribolium castaneum* (Wool, 1987).

Experimental design

Eight archipelagos, each with 18 islands, were colonized simultaneously. Each island was a vial with 1 g of standard medium, colonized with one pair of beetles. Each archipelago was arranged as a grid of six columns and three rows. The genotypes of the founding pairs were determined by their position in the grid. Islands in each column were colonized by a beetle from a strain homozygous for one of six recessive mutations. The islands in each row were colonized with a beetle (of the opposite sex) of one of three wild-type strains. Thus all the offspring in the first generation in all islands were heterozygotes, with the mutant allele frequencies identical at 0.5.

All adults were removed after a week of mating and oviposition, and the eggs were allowed to develop to adults, which were classified as mutant or wild type. Three pairs of siblings were selected randomly from among the offspring and placed in vials with medium. One fertile pair continued the 'island' population for the next generation (this produced a one-pair bottleneck every generation). The experiment lasted 11 generations.

Three archipelagos were held in a constant, optimal environment (IHC). Three others were given a series of random, deleterious treatments twice a week during the larval period (IHV). The last two archipelagos (OC, OV) started as IHC, but from generation 5 migration was enforced among them by interchanging the wild-type, but not the mutant, mating partner.

Gene diversity

Every generation the frequency q_i of the mutant marker among the offspring was estimated as $\sqrt{Q}$ (where Q is the observed proportion of mutant individuals). Gene diversity was estimated by the Shannon information index, where p_i is the frequency of the mutant allele in population i and

$$H = -\Sigma p_i \log_2 p_i$$

Total diversity was partitioned into among-island and within-island components (Lewontin, 1972).

Frequency distribution of alleles

The initially near-normal frequency distribution of the mutant allele among islands became U-shaped (see Chapter 4), as more and more populations became fixed at allele frequencies 0.0 or 1.0 (see also Wright and Kerr, 1954). Fixation was not symmetrical, however, as more populations lost the mutant than the wild-type allele, indicating that the mutations were in some way deleterious (Fig. 17.7).

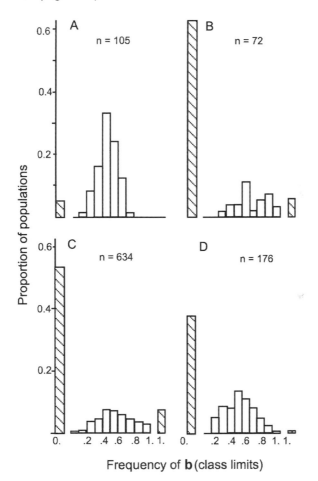

Fig. 17.7 Allele frequency distributions in IHV and IHC during the colonization process (from Wool, 1987; reprinted with permission of University of Chicago Press). A: generation 2; B: generation 11. C: all replicates pooled, generation 5-11. D: OV, generations 6-11. Striped bars: replicates fixed at $q_b = 0$ or 1.

Diversity within islands decreased in the inbred archipelagos, IHC and IHV, as expected from sib mating (calculated from Haldane's recurrence equations, Chapter 7). When outbreeding was enforced at generation 5, diversity quickly increased and stabilized in OC and OV at twice the level of the inbred lines (Fig. 17.8).

Diversity among islands increased steadily in the inbred archipelagos in the first four generations and stabilized at 60% of total diversity (Fig. 17.9). The island populations were becoming more and more genetically different from each other. When outbreeding was enforced in OC and OV, diversity among islands in these archipelagos decreased quickly (Fig. 17.9).

IC and IV changed at a similar rate and showed the same general temporal pattern. The mating system (inbreeding, with narrow bottlenecks)

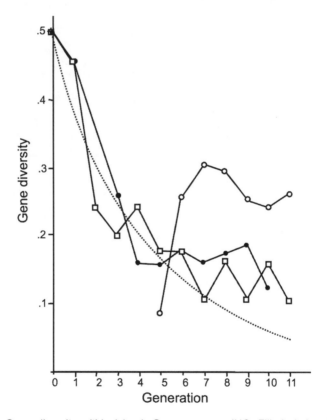

Fig. 17.8 Gene diversity <u>within</u> islands.Open squares: IHC; Filled circles: IHV. Open circles: OV. The smooth line is the expected heterozygosity in sib-mating populations, calculated from Haldane's recurrence equations. (From Wool, 1987; reprinted with permission of University of Chicago Press.)

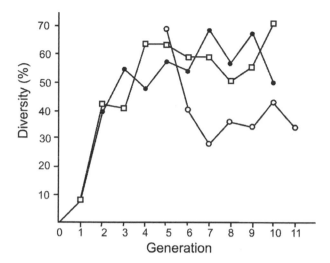

Fig. 17.9 Percentage among-island diversity in the experimental archipelagos. Open squares: IHC; Solid circles:IHV; Open circles: OV. Among-island diversity changed similarly in the two environments, affected mainly by inbreeding and bottlenecks. (From Wool, 1987; Reprinted with permission of University of Chicago Press).

seemed to be more effective than the external environment in the differentiation of the island populations.

Comparisons with real island populations

Eight sets of data from the literature on *Drosophila*, mice, and lizards in real archipelagos were consulted (Table 2 in Wool, 1987). Calculations of diversity from allele frequencies in the real data showed that diversity among island populations of flying organisms (*Drosophila*) was lower than in the model islands, while the populations of lizards were even more extremely differentiated than the model islands.

The experiment illustrates that a model gives, in a relatively short time, results qualitatively analogous to a process which in nature may be impossible to study.

Literature Cited

Ayala, F.J., J.R. Powell and T. Dobzhansky. 1971. Polymorphism in continental and island populations of *Drosophila willistoni*. Proceedings of the National Academy of Science, USA, 68: 2480-2483.

Bryant, E.H. 1969. The fates of immatures in mixtures of two housefly strains. Ecology 50: 1049-1069.

Gorman, G.C., M. Soule', S.Y. Yang and E. Nevo. 1975. Evolutionary genetics of insular Adriatic lizards. Evolution 29: 52-71.

Haj-Ahmad, Y. and D. Hickey. 1982. A molecular explanation of frequency-dependent selection in *Drosophila*. Nature 299: 350-352.

Lewontin, R.C. 1955. The effects of population density and composition on viability of *Drosophila melanogaster*. Evolution 9: 27-41.

Lewontin, R.C. 1966. Is nature probable of capricious? BioScience 16: 25-27.

Lewontin, R.C. 1972. The apportionment of human diversity. Evolutionary Biology 6: 381-398.

Sokal, R.R. and I. Huber. 1963. Competition among genotypes in *Tribolium castaneum* at varying densities and gene frequencies (the *sooty* locus). American Naturalist 47: 169-184.

Sokal, R.R. and F.J. Sonleitner. 1968. The ecology of selection in hybrid populations of *Tribolium castaneum*. Ecological Monographs 38: 345-379.

Stam, P. 1975. Linkage disequilibrium causing selection at a neutral locus in pooled *Tribolium* populations. Heredity 34: 29-38.

Wallace, B. 1969. Topics in Population Genetics. Norton, New York, USA.

Wool, D. 1970. Deviations of zygotic frequencies from expectation in eggs of *Tribolium castaneum*. Genetics 66: 115-132.

Wool, D. 1976. Changes of life history stage distribution of single strain and mixed *Tribolium* populations during a single generation at a lowered temperature. Journal of Animal Ecology 45: 381-394.

Wool, D. and E. Sverdlov. 1976. Sib-mating populations in an unpredictable environment: Effects on components of fitness. Evolution 30: 119-129.

Wool, D. 1987. Differentiation of island populations: a laboratory model. American Naturalist 129: 188-202.

Wright, S. and W.E. Kerr. 1954. Experimental studies of the distribution of gene frequencies in very small populations of *Drosophila melanogaster*. II. Bar. Evolution 8: 225-240.

CHAPTER 18

The Neutralist–Selectionist Controversy: 'Non-Darwinian' Evolution?

The present chapter was written when the controversy over the question of selective neutrality of mutations was raging, and molecular evolution was in its infancy. In retrospect, the balance of opinion moved in favor of the neutralists. But even had the neutrality theory been found wrong, it certainly was beneficial to science and to evolutionary theory – as Darwin had claimed,

> False facts are highly injurious to the progress of science, for they often endure long. But false views, if supported by some evidence, do little harm, for everyone takes salutary pleasure in proving their falseness. And when this is done, one path towards error is closed, and the road to truth is often at the same time opened (Darwin, 1874, p. 630).

HISTORICAL REVIEW

Do all characters evolve via natural selection? Darwin himself had some doubts, and allowed that neutral characters may exist:

> I am convinced that … many characters which may now appear to be useless will hereafter be proved to be useful, and will therefore come within the range of natural selection. Never the less, I did not formerly consider sufficiently the existence of structures which, as far as we can at present judge, are neither beneficial nor injurious. And this I believe to be one of the greatest oversights as yet detected in my work (Darwin, 1874, p. 76).

One of the foremost evolutionists (= selectionists) of the 20th century, Ernst Mayr, dedicated a whole section in his book in 1942 to 'Neutral Polymorphism'. He presented the arguments for neutral variation (in morphological characters) very clearly:

> There is no reason to believe that the presence or absence of a band on the snail's shell would be a noticeable selective advantage or disadvantage (Mayr, 1942, p. 75).
>
> It should not be assumed that all the differences between populations and species are purely adaptational and that they owe their existence to their superior selective qualities. Many combinations of color patterns, spots and bands, as well as extra bristles and wing veins, are probably largely accidental (Mayr, 1942, p. 86).

Richard Goldschmidt – a refugee scientist from Nazi Germany who settled in California as a professor of biology – argued in 1933 that the accumulation of neutral mutations could facilitate speciation. In developing the concept of **pre-adaptation**, Goldschmidt suggested that neutral mutations could accumulate in individuals without causing damage. If and when conditions changed, these 'hopeful monsters' could turn out to be the best adapted for the new environment (Goldschmidt, 1933). Goldschmidt's views were rejected by his contemporaries and he was almost (scientifically) outlawed for expressing them.

THE CONTROVERSY

The controversy emerged from the observation of electrophoretic variation. Electrophoresis was a breakthrough in the search for genetic variation in populations, but there were difficulties in its interpretation. Too many genetic markers (bands on a gel) were discovered in almost every population and species, and the detectable differences between individuals (in migration distance on a gel) were too small to believe that they affected the fitness of their carriers, and that their frequencies in the population were maintained by natural selection. The discovery of a great many variants in natural populations raised the question: Does every individual with a different mobility of an electrophoretically detectable protein on a gel, represent a genotype with a different fitness?

The controversy broke out with the publication of a paper with the provocative title 'Non-Darwinian evolution' (King and Jukes, 1969). The authors supported M. Kimura's suggestion, published just one year before – that the majority of the detectable variation has no effect on fitness and no selective value whatsoever: it is selectively <u>neutral</u>.

Hundreds of research projects were designed to disprove this suggestion, and thousands of scientific publications appeared to support one side or the other of the argument. The biological world was widely split into two camps – neutralists and selectionists. The split remained for 20 years.

Some leading biologists reacted strongly against the possibility that electrophoretic variation may not be under selective control – perhaps because electrophoretic variation seemed so widespread that putting it beyond selection would undermine the role of selection in nature entirely.

MODELS OF THE STRUCTURE OF THE GENOME

Two conceptual models of the structure of the individual genome were offered, and historically affected evolutionary thought. The first, 'typological' concept, assumed that a diploid individual is homozygous at all sites except very few mutant heterozygous loci. A pair of homologous chromosomes may look like this,

$$...+ + + + + + m +...$$
$$...+ m + + + + + + ...$$

This model implied that natural selection may eventually remove the mutant alleles (most of which are deleterious) and preserve the 'wild-type' genome. Heterozygous loci are few, depending largely on mutation frequency.

The second model – called the neo-Darwinian model – assumed that most loci in an individual are in the heterozygous state – and only rarely homozygous. As each individual is genetically different from any other, the number of alleles at a locus is virtually infinite. Two homologous chromosomes in an individual may look like this:

$$...A_1 B_5 C_4 D_2 E_8 F_4 G_1 H_6 ...$$
$$...A_3 B_5 C_6 D_1 E_5 F_5 G_8 H_4...$$

The neo-Darwinian model assumes that populations are very variable. Every mutation has some selective value, and the frequencies of the mutants in populations are maintained largely by balancing selection. It is this model of genetic structure which was challenged by the neutrality hypothesis.

NEUTRALISM VERSUS NEO-DARWINISM: BASIC ASSUMPTIONS

The neo-Darwinian model of evolution is based on the following assumptions:

(1) Genetic variation among individuals is extensive, and is related to the individual fitness.

(2) Most mutations are deleterious. A few are favorable to the individual. Mutations that currently seem to have no obvious effects on the survival or reproductive success of their carriers, probably had such an effect in the past.

(3) Natural selection is the main driving force in evolution. The differential survival and reproductive success of genotypes gradually changes allele frequencies.

The neutrality hypothesis is based on the following assumptions (Nei, 1975):

(1) Considerable variation exists in populations independently of balancing selection (this variation is phenotypically expressed, in part, in electrophoretic mobility of proteins).

(2) This variation represents two types of mutations: (a) neutral, or equivalent, mutations, where the replacement of one allele by another has no effects on individual fitness, and (b), deleterious mutations en route to extinction (or, rarely, positive mutations en route to fixation).

The frequencies of the first kind of mutation are determined by genetic drift and random fixation. In infinitely-large populations the frequencies of the second type are determined by natural selection, but in real – usually small – populations, genetic drift and random fixation affect their frequencies more than natural selection.

(3) On average, per locus, $2Nu$ mutations are expected (where u is the mutation rate). $1/2N$ of them will reach fixation per generation.

(4) The time needed for fixation of a positively-selected mutation in a population of size N is 4N generations. In a large population this is a very long time indeed – e.g., if N = 10,000, fixation will require 40,000 generations.

(5) When a population is balanced, i.e., when the arrival of new neutral mutations equals the extinction of old ones, the mean heterozygosity in the population will be given by

$$H = 4Nu/(4Nu + 1)$$

(6) The main role of natural selection is to remove less-fit genotypes – not to preserve balanced polymorphisms.

(7) Finally, at the DNA level, evolution takes place mainly by the substitution of one allele by another (mutation) and not by changes of genotype frequencies via selection.

Neutralists maintained that the observed genetic variation at the molecular level is neither beneficial nor harmful – it is simply a historical record of random and selective processes which had occurred during evolution. It is unclear what – if any – is the value of this variation for evolution in the future (Nei, 1982).

Variation in populations is maintained quantitatively by the mutation rate. Mutations, although of rare occurrence, may be very important for evolution when genetic variation is greatly reduced (as after a bottleneck), and the population is unable to adapt to new conditions in the absence of suitable genotypes.

EVIDENCE IN SUPPORT OF THE NEUTRALITY HYPOTHESIS

(1) The degeneracy of the genetic code

Natural selection cannot detect changes in the DNA which are not expressed in phenotypes. Therefore mutations that result in synonymous codons – and do not alter the amino acid coded for in the protein – will not have an effect on fitness. The codons for the 20 most common amino acids were already known almost 50 years ago (Strickberger, 1968). It was noted that there is more than one code for some amino acids (a total of 61 of the 64 possible 3-letter codes ($4^3 = 64$) code for the 20 amino acids. The rest are termination codons). For example, CGG, CGA, CGU and CGC all code for the same amino acid, arginine (see above, Table 5.1).

It is possible to get nine potential codons from each of the 61 listed in the table by point mutations, for a total of 549 possible codons (61×9). Of these, 134 (24.4%) are synonymous. No change in the protein will occur when these mutated codons are translated. As far as natural selection is concerned, these synonymous changes must be neutral.

(2) Charge changes

Not all the non-synonymous substitutions are detectable by electrophoresis. Allelic changes in proteins, reflected in gels by migration of allozymes to different distances on the gel, are detectable only if there was a charge change in the protein as a whole (for example, a point mutation changing the codon AGA (arginine) to AGU (serine) will remove one positive charge from the protein). Not all substitutions in the DNA result in a charge change in the protein.

Calculations by Nei (1975) show that of 392 possible non-synonymous nucleotide substitutions in the genetic code, only about 1/3 cause an electric charge change (Table 18.1): two thirds of the substitutions do not cause a charge change, and are undetectable by electrophoresis.

(3) Comparisons of homologous proteins from different organisms

Homologous proteins fulfil the same function in different organisms, and are similar in their molecular structure. They are probably descended from

Table 18.1 Frequency of protein charge changes due to mutations in the genetic code (from Nei, 1975)

Amino-acid substitution	Charge change	Percentage of cases
Neutral to positive	+1	10.72
Negative to positive	+2	0.51
Neutral to negative	−1	5.10
Positive to negative	−2	0.51
Positive to neutral	−1	10.72
Negative to neutral	+1	5.10
Total		32.66
No charge change	0	67.34

a single ancestral protein. The amino-acid sequences of many proteins are stored in data banks. One of the first sequences to be published was that of Cytochrome C, a co-enzyme in the respiratory metabolic system. Sequences of Cytochrome C from 38 organisms, ranging from yeast to man, were compared (Dickerson, 1972). This molecule has a heme active center and a chain of 104 amino acids. The heme group, and 35 of the 104 sites in the chain, were occupied by the same amino acids in all 38 organisms. At the other 69 sites, two to nine different amino acids occupied each site in different organisms – with no detectable change in the function of the molecule.

The interpretation of these data is that the heme, and the 35 fixed sites, are critical for the structure and/or the functioning of the molecule and therefore were conserved in evolution. The other 69 sites are free to vary – and substitution of amino acids may occur with no harmful effects. These substitutions do not alter the fitness of the individuals that carry them and are therefore selectively neutral.

A comparison of the numbers of substitutions in homologous proteins from different organisms – assuming that they are descended from the same ancestor – revealed a correlation between the numbers of substitutions per protein and the time that passed from their point of divergence (estimated from the phylogenetic tree of the organisms) (Kimura and Ohta, 1971; Ohta, 1974). Such a correlation is expected if most of these substitutions were neutral, and their accumulation depended only on mutation rate. Some proteins were evolving more slowly than others: this was interpreted as indication for different mutation rates at different sites.

If these substitutions were under selective control, organisms which develop faster and have more generations per unit time should have accumulated more substitutions. This was not the case: No correlation was found between the numbers of substitutions and the generation time of the organisms. The rate of accumulation of substitutions appeared to be constant for each protein, and variable among proteins, regardless of generation time. These empirical observations generated the '**molecular clock**' **hypothesis** which enables estimates of the time since divergence of two lineages from a common ancestor.

Literature Cited

Darwin, C. 1874. The Descent of Man and Selection in Relation to Sex. 2nd ed. Hurst & Co., New York, USA.

Dickerson, R.E. 1972. The structure and history of an ancient protein. Scientific American 226: 58-72.

Goldschmidt, R. 1933. Some aspects of evolution. Science 78: 539-547.

King, J.L. and T.H. Jukes. 1969. Non-Darwinian evolution. Science 164: 788-798.

Kimura, M. and T. Ohta. 1971. Protein polymorphism as a phase of molecular evolution. Nature 229: 467-469.

Mayr, E. 1942. Systematics and the Origin of Species. Columbia University Press, New York, USA.

Nei, M. 1975. Molecular Population Genetics and Evolution. North-Holland, Amsterdam, The Netherlands.

Nei, M. 1982. Genetic polymorphism and the role of mutations in evolution. pp.165-190. In: M. Nei and R.K. Koehn (eds.), Evolution of Genes and Proteins, Sinauer, Sunderland, USA.

Ohta, T. 1974. Mutational pressure as the main cause of molecular evolution and polymorphism. Nature 252: 351-354.

Strickberger, M.W. 1968. Genetics. MacMillan, New York, USA.

CHAPTER 19

The Neutrality Hypothesis: Molecular Support – and Evidence to the Contrary

By the end of the 20th century most evolutionists accepted that non-selective processes – genetic drift and random fixation – are responsible for evolutionary change at the molecular level, and that most of the mutational changes in the DNA are neutral or slightly deleterious (many of the latter are eliminated early by natural selection).

GENE DUPLICATION AND PSEUDOGENES

One mechanism which enables the accumulation of 'small' genetic changes – and can in time cause major evolutionary alterations (the hypothetical 'hopeful monsters' of Goldschmidt, 1933) – is **gene duplication**. If the coding sequence for some important gene is duplicated, so that the genome includes two (or more) copies of that gene, one of the copies may accumulate all kinds of sequence changes without disrupting the phenotypic expression, so long as the other copy remains unchanged and functions properly. An example of gene duplication was recently discovered in a study of embryological development of the flour beetle *Tribolium castaneum.* The two gene copies completely diverged from each other. Each duplicate acquired a different expression pattern and a separate function (Van der Zee et al., 2005). If one gene copy is inactivated, the mutated, inactive gene is termed a **pseudogene** (Li, 1983).

Pseudogenes are similar in sequence to the original active genes but are inactivated by some deleterious mutation. Neutral mutations can be detected by comparing the sequences of active genes with pseudogenes (Graur and Li, 2000). Mutations in the pseudogene are neutral because the pseudogene is already inactive and is not expressed phenotypically, thus cannot be affected by natural selection.

In their paper entitled 'Selfish DNA', Orgel and Crick (1980) revived Goldschmidt's old idea in a new form: a DNA sequence that has no effect on the phenotype lies dormant in a cell and accumulates mutations, out of reach of natural selection (while dormant, it is selectively neutral) like a parasite adapted to its host. A situation may arise in which this mutated sequence ('pseudogene') will turn out to be the best adapted to some new environment, and will give the pseudogene an immediate advantage (Orgel and Crick, 1980).

Potential sites of neutral mutations occur in the nuclear genome. Molecular research indicates that DNA sequences which code for proteins (the **exons**) are separated by large sections of non-coding nuclear DNA (spacers, **introns**). The introns are not translated into proteins, and thus can accumulate mutations without affecting the phenotype. Moreover, large parts of the nuclear DNA are repetitive – many copies of the same sequence occur in the genome. The value of this mass of repetitive DNA to the organism is not clear, but mutations in these sequences may not be harmful so long as at least one copy remains functional. For more information on these matters see Li and Graur (1991) or Graur and Li (2000).

'Haldane's Cost'

J.B.S. Haldane calculated the numerical effect ('cost') that natural selection can impose on a population. Selection for a given genotype – a substitution of one allele by another – means that the less-favored genotypes (fitness $W_i < 1$) will die or leave fewer offspring. Haldane was interested in the ratio of the numbers of individuals destroyed by selection to the number surviving. In the case of selection favoring a recessive genotype, using the notation of chapter 9, the ratio is –

$$sq/q(1 - s)$$

After **t** generations, the cumulative cost will be

$$C = \Sigma^t(sq/q \ (1 - s))$$

Haldane called C 'the cost of gene substitution'. This cost is cumulative, and will be lower - the smaller is the selection intensity **s**. But unless **s** = 0, there will be a cost to selection (some of the reproductive effort of the population is wasted). If many independent alleles are under selection, Haldane claimed, the cost may exceed the reproductive output of the population! However, if mutations are selectively neutral (**s** = 0), there is no cost to the substitution of one allele by another. It is reasonable to suppose that most of the substitutions are in fact neutral.

[Haldane's argument was based on the assumption that the selected genes are independent of each other. This is not so: genes are linked on chromosomes. Moreover, the effect on an individual of carrying one or 100 lethal genes is the same: the individual dies.]

MODELS OF EVOLUTION BY NEUTRAL MUTATIONS: TESTING FOR NEUTRALITY

The most commonly-used model for evolution by neutral mutations is the **infinite-alleles model**, originally suggested as an explanation for the large amount of electrophoretic variation in natural populations.

The model applies the philosophical criterion called 'the principle of insufficient reason': if there is no particular reason to prefer a complex explanation, then the simplest explanation should be accepted. The simplest explanation for isozyme variation in migration distances on the gel is that this variation has no selective value. It is up to those who suppose otherwise to provide evidence in support of their view. In statistical terms, the neutrality hypothesis is a 'null' hypothesis. If the 'null' hypothesis is rejected, then a selective process can be assumed.

The infinite-number-of-alleles model assumes that there can be an infinite number of alleles at any locus. Random and neutral mutations occur rarely (at a rate **u** per site and generation). Each new mutation forms a new allele which has never existed before. In finite populations of (effective) size N, the probability that an individual is homozygous for allele **i** is

$$p_i^2 = 1 / (1 + 4 Nu)$$

(Crow and Kimura, 1970), where p_i is the frequency of that allele in the population. If **a** loci are sampled, the average homozygote frequency is expected to be $F = \Sigma\, p_i^2 / a$, and the frequency of heterozygotes is expected to be

$$H = 1 - F = 1 - 1/(4Nu + 1) = 4Nu/(4Nu + 1)$$

This formula can be used to test the validity of the neutrality hypothesis. The level of heterozygosity in natural populations can be measured electrophoretically. If the observed value exceeds the calculated

H, the neutrality hypothesis cannot be the proper explanation for this variation.

The difficulty is that no data are available on the magnitude of N (effective population size) or **u** (mutation rate) in natural populations. Crow and Kimura (1970) furnish a table of predicted values of H as a function of **u** and N (Crow and Kimura, 1970, Table 7.2.1). The table shows that almost all known levels of heterozygosity fall within reasonable boundaries of N and **u**, and will support the acceptance of the neutrality hypothesis.

Empirical Tests

Fuerst et al. (1977) collected published data on the average heterozygosity (and its variance) in natural populations of 95 species of vertebrates and 39 of invertebrates, in each of which at least 20 electrophoretic loci were sampled. They compared the observed level of heterozygosity with predictions from the infinite-allele and two other neutral models. The observed level of heterozygosity agreed with the neutral predictions from all three models (within the confidence limits).

Nei and Graur (1983) collected reliable estimates on natural population size and generation time of 75 organisms, together with data on electrophoretic variation (heterozygosity) in the same organisms. The neutrality hypothesis suggests that the mutation rate, **u** (mutations per site per year), is constant for each protein. Therefore the number of accumulated mutations per generation U is a function of **u** as well as the chronological length of a generation,

$$U = ug$$

[The observable number of differences between two proteins should depend on U and on the chronological time since the divergence of the compared proteins from their common ancestor.]

Nei and Graur (1983) calculated heterozygosity for their selected organisms from the relation $H = 1 - \Sigma p_i^2$, which is independent of the mating system and considered a more reliable estimate than counting of heterozygotes on the gels. If the neutralist hypothesis is correct, two predictions could be made: first, mean heterozygosity should increase with population size; and second, observed heterozygosity should be lower than or equal to the expected values from the expression

$$H = 4Nu/(4Nu + 1)$$

This is because the estimates of N (used in the formula) were probably too high. Bottlenecks probably occurred in the past history of the populations and heterozygosity is strongly affected by genetic bottlenecks

(Nei et al., 1975). If a population went through a bottleneck, then the level of heterozygosity in it should be lower than estimated using the current N.

Both predictions were supported by the calculations. H did increase with the (logarithm of) population size, and observed heterozygosity was lower than the expected values.

The reason why most mutational changes seem to have no selective value may be technical. It may be impossible to assign mutations to categories by their effects on fitness as favorable or deleterious: there is a continuum of mutation effects on fitness, from very deleterious to very favorable (selection coefficient **s** = −1 to +1) with most of them in the middle around the value of zero effect (true neutrality). It may not be possible to detect the very small effects on fitness of most mutations, and they will be 'effectively' neutral (Ohta, 1974).

SILENT SITES IN THE DNA

Sequences of DNA that are not translated into proteins (**silent sites**), like introns and pseudogenes, accumulate more mutations than active genes (**replacement sites**. Lewin 1983). Silent sites are 'effectively' neutral. As such, the frequency of accumulated mutations in pseudogenes and other silent sites should approximate the rate of spontaneous mutations. The observed frequency of mutations in active genes should be lower – because they can be deleterious and removed by natural selection.

In the table of the genetic code (Table 5.1) most of the single nucleotide substitutions at the <u>third</u> codon position are synonymous and should cause no change in the protein composition in active genes. An examination of real protein sequences shows in fact that many more replacements accumulated in the third codon position than in the first or second position (Li et al., 1981; Table 19.1). This pattern can be noticed even in pseudogenes, albeit to a smaller extent: since the pseudogenes are inactive, changes can accumulate also in the first and second positions (Table 19.2).

Li and Graur (1991) calculated the average rate of synonymous and non-synonymous mutations from data on mutation frequencies in many species of mammals. The average rate of synonymous mutations was 4.61 per 10^9 years, while non-synonymous mutations occurred at a much lower rate (0.85 per 10^9 years). This illustrates that synonymous mutations can accumulate (stay in the genome) longer than non-synonymous mutations, which may be removed by selection.

Table 19.1 Numbers of sequence differences in protein-coding genes at the three codon positions. Length = the number of codons in the sequence (from Li et al., 1981). Most of the substitutions at the third position are synonymous

Protein organisms compared		length	first position	second position	third position
α-globin	man-mouse	141	13	14	54.5
	mouse-rabbit	141	16	15	53
	man-rabbit	141	18	14	31
β-globin	man-rabbit	146	9	6.5	32
	mouse-rabbit	146	21.5	16	46

Table 19.2 Numbers of differences between pseudogenes (marked as ψ) and the corresponding active genes at the three codon positions (from Li et al., 1981)

Pseudogene	compared with	length	first position	second position	third position
Mouse ψ α-globin	Mouse α-globin	117-119	12	17	22
	Human α-globin		19	23	47
	Rabbit α-globin		21	22	48
Human ψ α-globin	Human α-globin	133	30	26	42.5
	Mouse α-globin		34	31	49
	Rabbit α-globin		39	32	51
Human ψ β-globin	Rabbit β-globin	145-146	26	19	34
	Human β-globin		31	24	42
	Mouse β-globin		34	31	43

ARE SYNONYMOUS SUBSTITUTIONS REALLY NEUTRAL?

One way to answer this question was to examine the frequencies of usage of alternative codes for the same amino acid. For example, there are eight amino acids which have four codons each, differing only in third codon position (GLY, ALA, VAL, THR, ARG, PRO, LEU, SER (Table 5.1). If synonymous substitutions are selectively neutral, then the four codons should be equally frequent in protein-coding genes. Berger (1977) checked that assumption in the mRNA which codes for the envelope proteins (1068 codons) of the virus MS2. The frequencies of the four codons were not equal. The frequencies of nucleotide pairs also were not equal: codons with A and T in third position were more frequent than codons with G and C.

This pattern was later discovered in other virus proteins and in other organisms. Gojobori et al. (1982) suggested that the A-T excess probably resulted from the chemical distance between the bases: the greater the chemical distance, the smaller is the substitution frequency between them. An extremely A-T rich content (80% A-T) was recently discovered in the DNA of the aphid symbiotic bacterium, *Buchnera aphidicola*. The authors examined several alternative explanations and concluded that this high content was maintained by a balanced mutation rate (AT → GC / GC → AT) and was not the result of balancing selection (Wernegreen and Funk, 2004).

GENOTYPE DISTRIBUTIONS IN NATURAL POPULATIONS: IS ELECTROPHORETIC VARIATION SELECTIVELY NEUTRAL?

If electrophoretic markers are selectively neutral, the frequencies of these markers in natural populations should vary randomly among sites. This is often not the case: the same alleles are the most frequent in sites very distant from each other – for example in humans (Lewontin and Krakauer, 1973) and in *Drosophila* (Ayala, 1974).

Ayala et al. (1971) examined the frequencies of some electrophoretic markers and chromosomal inversions, in samples of *Drosophila* from sites in South America and the Caribbean islands. The inversion polymorphism seemed to be strongly affected by chance: inversion frequencies were indeed very different from one island to the next, and all inversions found on the islands were also represented in samples from the South American mainland – as expected from founder effects and random processes. By contrast, the frequencies of electrophoretic alleles were similar in widely-distant islands. This is difficult to explain unless the electrophoretic variation is strongly controlled by natural selection.

Natural selection may affect genotypes at single loci, favoring or selecting against different genotypes at different locations. Genetic drift, and inbreeding caused by isolation by distance, must work on the entire genome. When individuals colonize different islands, the small founder gene pool should vary among islands. Inbreeding may further limit the number of genotypes and increase the differences among populations (Lewontin and Krakauer, 1973). Thus the variance among populations may be used to test if differences among them are due to selection.

The expected variance in frequency of a selectively neutral marker in a randomly-mating population is

$$\sigma^2 = pq$$

where **p** and **q** are the mean allele frequencies of all populations in the study. If the observed genetic variance is greater than this expectation, it means that random factors are insufficient to account for the differences, supporting the conclusion that the variation may be under selective control. Lewontin and Krakauer suggested that

$$f = s_p^2/pq$$

may be used as a test criterion (where s^2_p is the observed variance). They used human blood-group frequency data to show that differences in these frequencies among humans from different countries cannot be explained by chance alone.

SELECTION AFFECTING ELECTROPHORETIC VARIATION: FIELD STUDIES

Isozyme polymorphism (estimated as mean numbers of alleles per locus, or heterozygosity) in enzymes which participate in critically important functions of the body, like metabolism, was lower than in enzymes with no known critical functions (Powell, 1975; Ayala and Powell, 1972). This result was offered as indirect support for the claim that the frequency of these isozymes is under selective control.

Three genotypes at the EST-1 locus were found in natural populations of the fish *Catostomus clarki*. Research showed that the maximal activity of the enzyme extracted from the homozygote **a/a** was at a temperature of 37°C, declining at lower temperatures. The enzyme from the homozygote **b/b** was maximally active at 0°C, declining at elevated temperatures. The enzyme from the heterozygote **a/b** was active in a wide range of intermediate temperatures. These results fitted well with the observed distribution of the three genotypes along the eastern Atlantic shores of America, with **a/a** frequent in the south (Florida) and **b/b** in the north (Massachusetts) (Koehn, 1969).

Geographical Clines

Geographical or climatic clines in frequencies of electrophoretic markers were studied as indications of their regulation by natural selection. Frequencies of electrophoretic markers often varied along geographical gradients, like north-to-south, distance from the sea, or elevation. For example, Johnson (1976) plotted the frequencies of the fast (F) and slow-migrating (S) allele of the enzyme α-GPDH in samples of a butterfly (genus *Colias*) from different sites, against a measure of environmental predictability (estimated as the number of daylight hours in which the temperature was suitable for flight and reproduction). He found that at approximately 2000 m elevation, the environment was unstable and changed rapidly, while at higher and lower elevations the environment was relatively constant and predictable. Isozyme variation in the butterflies was much greater in the unpredictable zone than in the stable habitats (estimated heterozygosity H = 0.35 versus 0.06-0.07). Heterozygotes seemed to have an advantage when the environment was unpredictable.

The existence of electrophoretic variation in organisms from the deep sea (more than 2000m below the surface), which was considered a particularly stable environment, was used to claim that either this variation is independent of selection – or, contrariwise, that the deep-sea environment is not as stable as it seems to be (Gooch and Schopf, 1972).

Nevo (1978) showed that electrophoretic variation was higher in 117 species categorized as ecological 'generalists', than in 120 ecological 'specialists' (average numbers of alleles per locus, 0.348 versus 0.189; average heterozygosity, 0.106 versus 0.046). This showed that in species with more restricted requirements from the environment, natural selection tolerates a narrower range of variation.

LABORATORY STUDIES

Alcohol Dehydrogenase (ADH) in *Drosophila*

Alcohol is formed by yeast in fermentation in rotting fruit and in the wine industry. ADH hydrolyzes the alcohol into CO_2 and H_2O. Since alcohol is toxic to the yeast, as well as to the *Drosophila* larvae which feed in the fruit, the hydrolytic activity of ADH is beneficial to them. There are two principal forms of ADH detectable by electrophoresis, a Fast (F) and a Slow (S) allele. F activity is stronger than S.

Forty populations of *D. melanogaster* were formed by crossing homozygous ADH-F with ADH-S flies. The cross ensured that all populations were started at equal allele frequencies (q_F = 0.5). Five replicate populations each were reared on media containing one of seven alcohols (methanol, ethanol, propanol, isopropanol, butanol, hexanol and glycerol). One group of 5 replicates was reared on alcohol-free medium as control. The frequency of the F allele was measured in a sample of 150 flies of each group every generation for 10 generations.

On media containing alcohol, the frequency of ADH-F increased from 0.5 to 0.8 or even to 1.0. In the control and glycerol groups, the frequency of ADH-F did not change. The carriers of the stronger-acting F allele seemed to have a selective advantage when the medium contained alcohol: i.e., this variation was not neutral (Van Delden and Kamping, 1983).

An ADH-null strain is fully viable in the laboratory on media without alcohol, and even occurs in low frequencies in nature (van Delden, 1982). In the presence of ethanol in the medium (17% alcohol was the median dose), LD_{50} of flies of an ADH-F strain was six times as high as that of the

ADH-null strain. Adult *Drosophila* survived longer if held in an atmosphere containing 2% ethanol in water vapor, than in an atmosphere with water vapor alone. When ethanol concentration is too low to be harmful, the flies may be able to use ADH to break down the alcohol in the vapor – and use it as an additional energy source (Daly and Clarke, 1981).

Amylase Variation in *Drosophila*

Amylases (AMY) are secreted in the salivary glands, and in the fore- and midgut of insects. These enzymes digest starch to disaccharides, mostly to maltose, *in vitro* and *in vivo*.

In a comprehensive study of AMY activity in *Drosophila*, two strains homozygous for different AMY isozymes were compared: $Amy^{4,6}$ is very active, and Amy^1 has low activity (De Jong and Scharloo, 1976; Hoorn and Scharloo, 1978). There was no difference in survival of the two strains when reared in pure culture. When the quantity of yeast was reduced to a minimum, survival largely depended on the quantity of starch. When larvae of both strains were introduced together to the jars with variable quantities of starch (and yeast quantity held to the minimum), the frequency of $Amy^{4,6}$ increased with the quantities of starch in the medium.

This experiment demonstrated that AMY electrophoretic variation is not neutral: when starch was limiting, there was an advantage to the genotype which was more effective in hydrolyzing starch.

CONCLUDING REMARKS

The Darwinian and neutralist hypotheses seemed irreconcilable at the time. Twenty years of dispute – and of research to discover the truth – have brought the two sides a lot closer together. Darwinists are ready to accept that molecular processes are largely random. Neutralists are ready to agree that some selection is taking place on some characters, although natural selection has only a minor role in evolution.

Literature Cited

Ayala, F.J., J.R. Powell and T. Dobzhansky. 1971. Polymorphism in continental and island populations of *Drosophila willistoni*. Proceedings of the National Academy of Sciences, USA 68: 2480-2483.

Ayala, F.J. and J.R. Powell. 1972. Enzyme variability in the *Drosophila willistoni* group.VI. Levels of polymorphism and the physiological function of enzymes. Biochemical Genetics 7: 331-345.

Ayala, F.J. 1974. Biochemical evolution: natural selection or random walk? American Scientist 62: 692-701.

Berger, E.M. 1977. Are synonymous mutations adaptively neutral? American Naturalist 111: 606-607.

Crow, J.F. and M. Kimura. 1970. An Introduction to Population Genetics Theory. Harper & Row, New York, USA.

Daly, K. and B. Clarke. 1981. Selection associated with the alcohol dehydrogenase locus in *Drosophila melanogaster*: differential survival of adults maintained on low concentrations of ethanol. Heredity 46: 219-226.

De Jong, G. and W. Scharloo. 1976. Environmental determination of selective significance or neutrality of amylase variants in *Drosophila melanogaster*. Genetics 84: 77-94.

Fuerst, P.A., R. Chakraborty and M. Nei. 1977. Statistical studies on protein polymorphisms in natural populations. I. Distribution of single locus heterozygosity. Genetics 86: 445-483.

Gojobori, T., W-H. Li and D. Graur. 1982. Patterns of nucleotide substitutions in pseudogenes and functional genes. Journal of Molecular Evolution 18: 360-369.

Goldschmidt, R. 1933. Some aspects of evolution. Science 78: 539-547.

Gooch, J.L. and T.J.M. Schopf. 1972. Genetic variability in the deep sea: relation to environmental variability. Evolution 26: 545-552.

Graur, D. and W-H. Li. 2000. Fundamentals of Molecular Evolution. Sinauer, Sunderland, USA.

Hoorn, A.J.W. and W. Scharloo. 1978. The functional significance of amylase polymorphism in *Drosophila melanogaster*. I. properties of two amylase variants.Genetica 49: 173-180.

Johnson, G.B. 1976. Polymorphism and predictability at the α-glycerophosphate dehydrogenase locus in *Colias* butterflies: gradients of allele frequency within single populations. Biochemical Genetics 14: 403-426.

Koehn, R.K. 1969. Esterase heterogeneity: dynamics of a polymorphism. Science 163: 943-944.

Lewin, B. 1983. Genes. Wiley, New York, USA.

Lewontin, R.C. and J. Krakauer. 1973. Distribution of gene frequencies as a test of the theory of selective neutrality of polymorphisms. Genetics 74: 175-195.

Li, W-H. and D. Graur, 1991. Fundamentals of Molecular Evolution. Sinauer, Sunderland, USA.

Li, W-H., T. Gojobori and M. Nei. 1981. Pseudogenes as a paradigm of neutral evolution. Nature 292: 237-239.

Li, W-H. 1983. Evolution of duplicate genes and pseudogenes. pp.14-37. In: M. Nei and R.K. Koehn (eds.), Evolution of Genes and Proteins, Sinauer, Sunderland, USA.

Nei, M., T. Maruyama and R. Chakraborty. 1975. The bottleneck effect and genetic variability. Evolution 29: 1-10.

Nei, M. and D. Graur. 1983. Extent of protein polymorphism and the neutral mutations theory. Evolutionary Biology 17: 73-118.

Nevo, E. 1978. Genetic variation in natural populations – patterns and theory. Theoretical Population Biology 13: 121-177.

Ohta, T. 1974. Mutational pressure as a main cause of molecular evolution and polymorphism. Nature 252: 351-354.

Orgel, L.E. and F.H.C. Crick. 1980. Selfish DNA: the ultimate parasite. Nature 284: 604-607.

Powell, J.R. 1975. Protein variation in natural populations of animals. Evolutionary Biology 8: 79-119.

Strickberger, M.W. 1968. Genetics. MacMillan, New York, USA.

Van Delden, W. 1982. The alcohol dehydrogenase polymorphism in *Drosophila melanogaster* : selection at an enzyme locus. Evolutionary Biology 15: 187-222.

Van Delden, W. and A. Kamping. 1983. Adaptation to alcohols in relation to the alcohol dehydrogenase locus in *Drosophila melanogaster*. Entomologia Experimentalis et Applicata 33: 97-102.

Van der Zee, M., N. Berns and S. Roth. 2005. Distinct functions of the *Tribolium* zerknüllt genes in the serosa specification and dorsal closure. Current Biology 15: 624-636.

Wernegreen, J.J. and D.J. Funk. 2004. Mutation exposed: a neutral explanation for extreme base composition of an endosymbiont genome. Journal of Molecular Evolution 59: 849-858.

CHAPTER 20

Molecular Evolution

HISTORICAL NOTES

The term 'molecular evolution' – as used in the titles of books like that by Graur and Li (2000) – covers two different, though necessarily related, fields of research: the evolution of genomic DNA or its derivatives – the proteins; and the study of the evolution of organisms as inferred from molecular data.

Historically, the study of molecular evolution was made possible when entire sequences of some important proteins – like cytochrome C, insulin and the globins – became available and homologous proteins in different organisms could be compared. Due to the direct relationship between the amino-acid sequence and the DNA coding for it, through the genetic code, it was possible to reconstruct the DNA sequences. This step enabled quantitative estimation of similarity and a new look at phylogeny. The study of evolution at the molecular level took a jump forward when the methods of 'reading' DNA directly were improved and they can now be carried out automatically by machines.

DNA sequencing revealed similarities among different genes, suggesting that they could be arranged as **gene families** which diverged from common ancestors. The human globin gene family was one of the first to be recognized. It contains two gene clusters, α and β, which are located on different chromosomes. The α cluster contains three active genes and two pseudogenes. The β cluster usually contains five active genes and one pseudogene (Lewin, 1981, 1983). The active genes are separated by introns. Similar gene clusters were found in the globin gene family of other mammals. Comparisons of DNA sequences in gene families enable the tracing of their common ancestry (molecular phylogeny) and the reconstruction of the phylogeny of the organisms carrying the gene sequences. This became one of the major issues in evolutionary research in the late 20th century.

The molecular evidence changed the **concept of the gene**. The classical model of a chromosome was a linear arrangement of genes, like a 'string of beads'. This model was very useful for understanding genetic phenomena like recombination, inversion and duplication (e.g., Strickberger, 1968). A gene is conceived in this model as a linear string of letters (nucleotides), which are read in triplets in linear order and translated into proteins. However, this concept was found too simplistic, and has now been replaced by a complex model of active, translatable segments (**exons**) separated by inactive segments (**introns**), which are not translated or not even transcribed. The exons coding for a particular protein are not necessarily arranged side by side on the same chromosome: they are put together only during the transcription process, in the messenger RNA molecule (Lewin, 1983; Graur and Li, 2000).

The same exon may appear multiple times in the genome, on the same or different chromosomes. Of particular importance was the discovery of pseudogenes. Comparison of a pseudogene with its active gene is important in the context of the neutrality hypothesis (Chapter 19) as well as for the understanding of the causes of gene inactivation.

EVOLUTION OF MOLECULES: PROTEIN EVOLUTION

Kimura (1983) listed five principles for protein evolution:

(1) Each protein evolves at an approximately constant rate. The number of amino-acid substitutions per site per year remains constant so long as the function and the tertiary structure of the protein remain the same.

(2) Non-essential molecules or parts of molecules which are less important for their function, evolve faster (= accumulate more changes per unit time) than more important molecules.

(3) Mutations causing amino acid substitutions which are less damaging to the structure and function of the protein, will be found more frequently than mutations which cause heavy damage.

(4) Duplication of an older gene sequence must precede the appearance of a new sequence.

(5) Selective elimination of deleterious mutations, and random fixation of neutral ones, are much more frequent in evolution than fixation of beneficial mutations by natural selection.

One of the first molecular studies of protein evolution was the comprehensive study of Cytochrome C (Dickerson, 1972). This study indicated great similarity of the amino-acid sequence from 38 different organisms, from yeast to man (Table 20.1). This similarity supported the conclusion that all these organisms had a common ancestor in the remote past. The same amino acid occurred at 35 of 104 sites in the molecule in all 38 organisms: these sites were probably essential for the function of the protein. Extant proteins diverged from a small number of ancestral ones, because it is 'easier' (more likely) to change an existing protein without destroying its function, than to build a new protein from amino acids to fulfill that fuction (Doolittle, 1981).

Table 20.1 Frequency distribution of amino acid differences in the sequence of Cytochrome C from 38 organisms, from yeast to man. The molecule is a chain of 104 amino acids (from Dickerson, 1972)

Different amino acids per site	number of sites
1	35
2	23
3	17
4	12
5	9
6	8

Some spontaneous mutations may enable the utilization of a new substrate. This was the case in β-galactosidase in bacteria: If a new substrate is available, the mutant enzyme is able to use it when the original enzyme is not (Hall, 1982). No such cases of accepting new function by spontaneous mutants are known in multicellular organisms. Mutations, unless they are neutral, more often than not will disrupt the activity of the resulting protein. The more similar is the new to the old sequence, the greater the chance that the mutant protein will still fulfil the function.

Most investigators tend to agree with Kimura (1983) that the evolution of new proteins is preceded by gene duplication. This can happen in two ways: duplication of a single gene, when the duplicates remain side by side **(tandem duplication)** or duplication of the entire genome, resulting in a polyploid. A series of tandem duplications may have led to the evolution of collagen genes in chickens, where about 50 exons are involved. The numbers of nucleotides in the sequences of some of these exons are multiples of 9: two exons have 45 nucleotides, 12 exons have 54, four have 99, and three have 108. They may have been formed by duplication of a single ancestral sequence (Li, 1983; Li et al., 1981). Repeated DNA sequences presumably formed by duplication, some containing up to 100,000 copies (referred to as **DNA mini-satellites**) may constitute 60% of the total DNA in cells (Nei, 1975).

One advantage of gene duplication may be enabling the production of large amounts of protein (e.g., an enzyme) quickly in a 'crisis' situation. Studies of insecticide resistance in the aphid, *Myzus persicae*, and in some mosquitoes, showed that organophosphorous-resistant strains contain multiple copies of an esterase isozyme which sequesters the insecticide (Field et al., 1988; Devonshire and Field, 1991; Mouches et al., 1986; Rooker et al., 1996).

RATES OF PROTEIN EVOLUTION

The rate of molecular evolution was first calculated from the number of amino-acid replacement in proteins. Today it is calculated from comparisons of DNA sequences of homologous genes in different organisms, which are stored in data banks.

Sequence comparisons

The calculation of evolutionary rate is based on the fundamental assumption that the compared genes diverged from a common ancestor, and that the observed sequence differences between them are due to accumulated mutations. The similarity of two derived proteins (or genes) is expected to decrease with evolutionary time since their divergence. For example, the human α, β, and δ-globin are assumed to have diverged from a single ancestral sequence by gene duplication. The sequences of α- and β-globin are 41% similar, those of β and δ are 73% similar. Hence the separation of β- and δ-globin in humans is a more recent event than the separation of β- and α-globin.

Proteins are composed of 20 common amino acids. The probability of occurrence by chance of the same amino acid at the same locus in two proteins is $20 * (1/20)^2 = 1/20 = 0.05$. (Actually, the probability of chance similarity is higher, because the frequencies of different amino acids in proteins are not equal: glycine, alanine and leucine constitute about ¼ of the amino acids in many proteins, while tryptophan and histidine are rare. The distortion caused by this unequal distribution may not be great if long sequences are compared.)

The probability of random similarity between sequences of DNA is greater: Since DNA is built of just four kinds of nucleotides, the chance of random identity of nucleotides at a site is 25%!

Difficulties in estimating sequence similarity are caused by <u>deletions</u> and <u>insertions</u> in the compared sequences (Doolittle, 1981; see Graur and

Li, 2000 for a review of the problems). The alignment of two sequences and determination of their similarity are carried out today by sophisticated computer algorithms.

Estimation of Rates of Protein (Gene) Evolution

Two statistics are used to measure evolutionary rate: the mean number of amino-acid replacements (per site and generation), or the mean number of nucleotide replacements (derived from the first measure). The data source may look like the globin data in Table 20.2.

Table 20.2   The number of amino-acid differences of the α-globin chain between different vertebrates (above diagonal); See text for the meaning of 2λt (below diagonal)

	Man	Horse	Cow	Carp
Man	0	18	16	68
Horse	0.138	0	18	66
Cow	0.121	0.138	0	65
Carp	0.666	0.637	0.624	0

If each of two proteins evolved at a rate of λ (selectively neutral) substitutions per site per year, the value of λ can be estimated from the number of observed differences between the two sequences as follows ('Poisson-correction method'. Nei, 1975). If each protein accumulated λ changes, the average number of mutations accumulated in both proteins in t years since their divergence from the common ancestor is expected to be 2λt (changes may have occurred in one of the sequences or in both).

From the Poisson distribution (consult a book on statistics, e.g. Sokal and Rohlf, 1995) the probability that **no** difference is detected – when the average number of differences is 2λt – is

$$p = e^{-2\lambda t}$$

If the protein contains n amino acids, and i of them are the same in both sequences, the proportion of identical (unchanged) sites between the proteins, i/n, is

$$i/n = e^{-2\lambda t}$$

or
$$2\lambda t = -\ln i/n$$

If **t** is known, or can be estimated from palaeontological data, then the value of λ can be derived from the last expression.

EVOLUTION OF MOLECULES AND OF ORGANISMS

The dramatic progress in DNA sequencing, the well-publicized applications of molecular methodology in forensic science, the cloning of farm animals (the sheep Dolly, for example) and the Human Genome project, capture public imagination. Students may obtain the impression that evolutionary research at the organismic level is a thing of the past or even obsolete. This is incorrect. Molecular changes by mutation and recombination occur within the cells of single individuals. To be transmitted to the next generation – other than by cloning – the individual carrying the mutant DNA must mate and reproduce. The processes determining whether the molecular change will eventually be fixed or lost - operate on the individuals, not on the DNA. **Genes evolve because the organisms carrying them live or die.** Whether randomly or selectively, the organisms and not the molecules are responsible for molecular evolution!

Molecular phylogenies often do not agree with the classical phylogenetic trees based on morphological characters. Although some tend to think that molecular trees are necessarily closer to the truth than the morphological ones, this is not necessarily so. The use of molecular data does not, a priori, guarantee that these kinds of data are better. The two approaches use different sets of data and should be complementary.

EVOLUTION OF ORGANISMS: THE 'MOLECULAR CLOCK' HYPOTHESIS

If the rate of evolution of each protein (λ mutations per year) is constant, then evolution proceeds like a clock, and the sequence difference of the protein between two species may be used to estimate the time of their divergence from their common ancestor during phylogeny. An independent estimate of the time scale is needed for calibration. The accuracy of these estimates depends on whether the rate of accumulation of mutations is in fact constant.

In the Hawaiian Archipelago, a time scale is provided by geological evidence. Carson (1976) incorporated the geological time scale on the age of the islands with his study of the colonization of the islands by Hawaiian *Drosophila*. Using electrophoretic markers, he found that the genetic distance between the endemic species was smaller in 'young' than in older islands (i.e., the species were more similar genetically to each other in the younger islands). He found a significant positive correlation between island age and genetic distance, and calculated that the rate of evolution (divergence of endemic species) was about 1% accumulated difference per 20,000 years.

To know whether both of the proteins changed or only one, the ancestral protein sequence must be known. This information may be acquired by assuming that sequences from more primitive organisms (as documented in their phylogeny) are ancestral to sequences from more 'advanced' organisms – but this need not necessarily be the case. (Goodman et al. (1982) claimed that the rates of protein sequence evolution were not constant, but were very fast after each duplication event and slowed down later. Their approach was heavily criticized by Kimura (1983).)

Another method of estimation of evolutionary rates – the **relative rate test** – is possible if the phylogenetic relationship of the studied organisms is known (or reliably inferred) (Figure 20.1). If species A and C had a common ancestor, the sequence divergence between them can be attributed only to the difference in evolutionary rates in the two lineages, since the time since divergence is the same. When the ancestral sequence is unknown, a third species B – known to be derived later from the A lineage – is used as a reference point (Li and Graur, 1991). If the molecular clock is true and the rate of nucleotide replacement is constant, then the number of differences between A and C should be the same as between B and C.

A comparison of 14 genes between mouse and rat showed that the evolutionary rates in the lineages of these two rodents did not differ, but a similar comparison between man and rat (representing the lineages of rodents and primates) showed that the rodent genes had evolved 4-6 times faster than the human genes (Li and Graur, 1991).

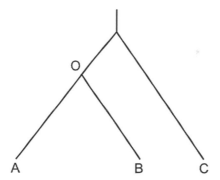

Fig. 20.1 Hypothetical phylogeny for the relative rate test. A, B and C represent observed sequences. The molecular clock hypothesis predicts that differences accumulated since sequences A and C diverged from a common ancestor should be the same as between B and C. Any deviation from equality must be attributed to unequal evolutionary rates in the two lineages, since the time since divergence is the same.

STUDIES ON MITOCHONDRIAL DNA

Many of the early molecular evolutionary studies used mitochondrial DNA (mtDNA). This is a compact molecule small enough to be packed into a plasmid or a bacterium, which was then the only way to clone it and obtain large enough quantities for analysis. This laborious procedure gave way to the much quicker – and now dominant – method of PCR, which also increased the limit size of analyzable sequences.

The use of mitochondrial DNA in population genetics and evolutionary studies has some advantages. This relatively small, circular molecule has a similar size (15700–19000 bp) in organisms from diverse groups (*Drosophila*, sea urchins, amphibians, reptiles, mouse and man). The complete 16,000 bp sequence of mtDNA of the honeybee is an example (Crozier and Crozier, 1993). In contrast with nuclear DNA, the entire mitochondrial genome is translated – no introns were found (Brown, 1983). It seems to code for the same genes in different organisms – 22 genes for t-RNA, two genes for r-RNA, and five protein genes. Research showed that while differences between individuals within species were in the order of 0.3-4.0% of the sequence, differences between mouse and man were about 30% of the sequence (Avise and Lansman, 1983). There is evidence that mtDNA is transmitted maternally, as a unit, with no recombination. Furthermore, about 1000 copies of mtDNA are found in every cell – ample material for comparative analysis of differences among populations within species. The fact that so many copies are present in every cell is perhaps the reason why no mtDNA repair mechanisms were discovered: inactivation of a copy does not cause major harm to the cell (Brown, 1983).

One study with mtDNA excited great public interest. Cann et al. (1987) compared mtDNA from the placentas of 145 human births, representing different nations from all continents. Their conclusion was that all humans are descendants of a single woman, who lived about 200,000 years ago. They also concluded that African human populations were older than (preceded) the European populations, a conclusion which agrees with the accepted African origin of man.

These conclusions were heavily criticized. The hypothetical ancestor was nicknamed 'Eve' (Wainscoat, 1987) and someone added sarcastically that if she really existed 200,000 years ago, she might have met Adam. Avise (1986) showed that even if all mankind has the same maternal mtDNA, this does not prove that only a single female carried this genome in the remote past: it could be shared by more than one woman (Latorre et al., 1986). In particular, it should be noted that by far most of the human genome – the nuclear genome – is not inherited maternally, and that crosses between different individuals from different populations created the great genetic diversity among humans of today.

The maternal inheritance without recombination of mtDNA gives the use of mtDNA an advantage in studies of descent of closely-related organisms on the phylogenetic scale (Avise, 1986). Comparative studies have shown that the rate of evolution of mtDNA (accumulation of mutations per unit time) is much faster than in nuclear genes in animals (not so in plants. Mitton, 1994). (The results of studies of this sort do not always agree with electrophoretic comparisons of the same organisms: for example, populations of the crustacean horseshoe crab *Limulus*, a 'living fossil', were electrophoretically identical, but mtDNA analysis discovered five distinct groups of these extant crabs.)

Molecular Clock for mtDNA Evolution

Brower (1994) suggested that mtDNA divergence occurred at a constant rate. Compiling data from different studies by other authors, a linear relationship was demonstrated between the percentage of sequence divergence between clades and the estimated age of the clades, yielding a constant divergence rate (regression slope) of 2.3% sequence divergence per million years. This estimate was used to relate the ages of divergence of populations of the grape phylloxera, *Dactulosphaira vitifoliae*, in the deserts of the southwestern USA to geological phenomena (Downie, 2004).

REGULATORY GENES IN EVOLUTION

Discussions of molecular evolution mostly deal with structural genes, which are translated into proteins. Some scientists assign a major role in evolution to genes which do not code for proteins, but regulate the activity of other genes that do – in particular in the processes of adaptation to new environments. For example, genes that control the amounts of enzymes produced in a metabolic pathway in stress situations (e.g. Hedrick and McDonald, 1980; Hall, 1982).

There is little hard, direct evidence of the adaptive value of regulatory genes, but there is circumstantial evidence in this direction (McIntyre, 1982). For example, man and chimp are extremely similar genetically – in some gene sequences they are identical – but morphologically and behaviorally they are very different. Perhaps these differences are due not to structural but to regulatory genes operating during ontogeny. Two parthenogenetic lines of *Drosophila mercatorum* are identical in 17 electrophoretic markers, but differ in morphology and other traits. The similarity in structural genes may not be the key to the understanding of the evolution of this species (Templeton, 1979).

Experiments on adaptation of *Drosophila* to media containing increasing concentrations of alcohols showed that after 28 generations of exposure, flies that were reared on these media contained a larger concentration of the enzyme ADH than control flies. There was no difference in the structural gene that codes for ADH nor in the biochemistry of its activity (McDonald et al., 1977). Comparisons of levels of ADH activity in two lines of *Drosophila* led to the conclusion that their third chromosome contained a regulatory gene, which affected the amount of enzyme produced by the structural ADH gene on the second chromosome (McDonald and Ayala, 1978). Controlled crosses with marked laboratory strains were used to map this regulatory gene.

A regulatory gene may be important when large amounts of an enzyme are required at short notice (Hedrick and McDonald, 1980). For example, Devonshire (1977) reported that the amount of the enzyme carboxylesterase in the aphid *Myzus persicae* increased 60-fold after exposure to a pesticide.

THE EVOLUTION OF THE GENETIC CODE

At some point in the evolution of life, self-replicating molecules must have appeared in the 'primeval soup'. The prevalence of DNA and RNA in extant organisms and their role in transmission of characters is proof that their formation was prior to the divergence of animal and plant groups. The 'universal' genetic code (Table 5.1) operates in all groups of the living world, from bacteria to man. How did this elaborate code evolve?

It is agreed that the primitive code was simpler than the present one. The fact that most synonymous substitutions occur in the third codon position, suggests that code evolution proceeded in the direction of reducing translational errors (Strickberger, 1968, p. 809).

Jukes (1966) called attention to the tRNA code, which is complementary to the messenger (mRNA). The 'anticodon' on the tRNA enables the recognition of the sites on the mRNA molecule and the attachment of the correct amino acid to its appropriate place in the protein molecure. Jukes suggested that the more primitive tRNA 'anticodons' recognized only 15 amino acids (Table 20.3). (This code was built of 16 codons, but one of them was not translated). Each of these 15 codons was able to recognize four amino acids, and all base substitutions at the 3rd position were synonymous.

A later stage in code evolution was the substitution of U by G at the first position, in some of these triplets. Each altered triplet now recognized only two amino acids. This reduced the ambiguity of the former codons as well as translational errors – and enabled the incorporation of new amino acids, increasing their number from 15 to 20. Supporting evidence for this scenario comes from the tRNA code in the mitochondria of mammals (Table 20.4. Jukes, 1983).

Table 20.3 Jukes' suggestion for the primitive anticodon. All 15 translated codons had U in first position. Each primitive triplet coded for 4 amino acids

Second position:	A	G	U	C
	UAA	UGA	(GUA)	ACU
	UAG	UGG	UUG	UCG
	UAU	UGU	UUU	UCU
	UAC	UGC	UUC	UCC

Table 20.4 The mitochondrial code in mammals (anticodon). G or U in first position. In parentheses, the amino-acid recognized by each triplet. * unaltered triplets (Table 20.3), each recognizing four mRNA codes for the same amino acid

Second position:	A	G	U	C
	UAA (leu)	UGA (ser)*	–	UCA (trp)
	GAA (phe)	*	GUA (tyr)	GCA (cys)
	UAG (leu)*	UGG (pro) *	UUG (glu)	UCG (arg)*
	*	*	GUG (his)	*
	UAU (met)	UGU (thr)*	UUU (lys)	–
	GAU (ile)	*	GUU (asn)	GCU (ser)
	UAC (val)*	UGC (ala)*	UUC (glu)	UCC (gly)*
	*	*	UGC (asp)	*

Literature Cited

Avise, J.C. 1986. Mitochondrial DNA and the evolutionary genetics of higher animals. Philosophical Transactions of the Royal Society London B 312: 325-342.

Avise, J.C. and R.A. Lansman. 1983. Polymorphism of mitochondrial DNA in populations of higher animals. pp. 147-164. In: M. Nei and R.K. Koehn (eds.), Evolution of Genes and Proteins, Sinauer, Sunderland, USA.

Brower, A.V.Z. 1994. Rapid morphological radiation and convergence among races of the butterfly *Heliconius erato* inferred from patterns of mitochondrial DNA evolution. Proceedings of the National Academy of Sciences, USA 91: 6491-6495.

Brown, W.M. 1983. Evolution of animal mitochondrial DNA. pp. 62-88. In: M. Nei and R.K. Koehn (eds.), Evolution of Genes and Proteins, Sinauer, Sunderland, USA.

Cann, R.L., M. Stoneking and A.C. Wilson. 1987. Mitochondrial DNA and human evolution. Nature 325: 31-36.

Carson, H.L. 1976. The unit of genetic change in adaptation and speciation. Annals of the Missouri Botanical Gardens. 63: 219-223.

Crozier, R.H. and Y.C. Crozier. 1993. The mitochondrial genome of the honeybee *Apis mellifera*: complete sequence and genome organization. Genetics 133: 97-117.

Devonshire, A.L. 1977. The properties of a carboxylesterase from the peach-potato aphid *Myzus persicae* Sulz. and its role in conferring insecticide resistance. Biochemical Journal 167: 675-683.

Devonshire, A.L. and L.M. Field. 1991. Gene amplification and insecticide resistance. Annual Review of Entomology 36: 1-23.

Dickerson, R.E. 1972.The structure and function of an ancient protein. Science 226: 58-72.

Doolittle, R.F. 1981. Similar amino acid sequences: chance or common ancestry? Science 214: 149-159.

Downie, D.A. 2004. Phylogeography in a galling insect, *Daktulosphaira vitifoliae* (Phylloxeridae) in the fragmented habitat of the southwest USA. Journal of Biogeography 31: 1754-1768.

Field, L.M., A.L. Devonshire and B.G. Forde. 1988. Molecular evidence that insecticide resistance in peach-potato aphid (*Myzus persicae* Sulzer) results from amplification of an esterase gene. Biochemical Journal 251: 309-312.

Goodman, M., M.L. Weiss and J. Czelusniak. 1982. Molecular evolution above the species level: branching pattern, rates and mechanisms. Systematic Zoology 31: 376-399.

Graur, D. and W-H. Li. 2000. Fundamentals of Molecular Evolution. Sinauer, Sunderland, USA.

Hall, B.G. 1982. Evolution on a petri dish: the evolved β-galactosidase system as a model for studying acquisitive evolution in the laboratory. Evolutionary Biology 15: 85-150.

Hedrick, P.W. and J.F. McDonald. 1980. Regulatory gene adaptation: an evolutionary model. Heredity 45: 83-97.

Jukes, T.H. 1966. Molecules and Evolution. Columbia University Press, New York, USA.

Jukes, T.H. 1983. Evolution of the amino acid code. pp.191-207. In: M. Nei and R.K. Koehn (eds.), Evolution of Genes and Proteins, Sinauer, Sunderland, USA.

Kimura, M. 1983. The Neutral Theory of Molecular Evolution. Cambridge University Press, Cambridge, UK.

Latorre, A., A. Moya and F.J. Ayala. 1986. Evolution of mitochondrial DNA in *Drosophila subobscura*. Proceedings of the National Academy of Sciences, USA 83: 8649-8653.

Lewin, B. 1981. Evolutionary history written in globin genes. Science 214: 426-429.

Lewin, B. 1983. Genes. Wiley, New York, USA.

Li, W-H., T. Gojobori and M. Nei. 1981. Pseudogenes as a paradigm of neutral evolution. Nature 292: 237-239.

Li, W-H. 1983. Evolution of duplicate genes and pseudogenes. pp.14-37. In: M. Nei and R.K. Koehn (eds.), Evolution of Genes and Proteins, Sinauer, Sunderland, USA.

Li, W-H. and D. Graur. 1991. Fundamentals of Molecular Evolution. Sinauer, Sunderland, USA.

McDonald, J.F., G.K. Chambers, J. David and F.J. Ayala. 1977. Adaptive response due to changes in gene regulation: a study with *Drosophila*. Proceedings of the National Academy of Sciences, USA, 74: 4562-4566.

McDonald, J.F. and F.J. Ayala. 1978. Genetic and biochemical basis of enzyme activity variation in natural populations. I. Alcohol dehydrogenase in *Drosophila melanogaster*. Genetics 89: 371-388.

McIntyre, R.J. 1982. Regulatory genes and adaptation: past, present and future. Evolutionary Biology 15: 247-286.

Mitton, J.B. 1994. Molecular approaches to population biology. Annual Review of Ecology and Systematics 25: 45-69.

Mouches, C., N. Pasteur, J.B. Berge, O. Hyrien, M. Raymond, B.R. Saint Vincent, M. Silvestri and G.P. Georghiou. 1986. Amplification of an esterase gene is responsible for insecticide resistance in California *Culex* mosquito. Science 233: 778-780.

Nei, M. 1975. Molecular Population Genetics and Evolution. North-Holland, Amsterdam, The Netherlands

Rooker, S., T. Gillemond, J. Berge, N. Pasteur and M. Raymond. 1996. Co-amplification of esterase A and B genes as a single unit in *Culex pipiens* mosquitoes. Heredity 77: 555-561.

Sokal, R.R. and F.J. Rohlf. 1995. Biometry, 3rd ed. Freeman, New York, USA.

Strickberger, M.W. 1968. Genetics. MacMillan, New York, USA.

Templeton, A. 1979. The unit of selection in *Drosophila mercatorum*. II. Genetic revolution and the origin of coadapted genomes in parthenogeneic strains. Genetics 92: 1265-1282.

Wainscoat, J. 1987. Out of the garden of eden. Nature 325: 13.

PART III

Macro-evolution

CHAPTER 21

The Concepts of 'Species' in Evolution

At the very outset of the inquiry, we are met with the difficulty of defining what we mean by the terms "species" and "race". (Lyell, 1853, p. 660).

The following may, perhaps, be reconcilable with the known facts. Each species may have had its origin in a single pair, or individual where an individual was sufficient. And species may have been created in succession, at such times and such places as to enable them to multiply and endure for an appointed period, and occupy an appointed space on the globe (ibid., p. 665)

MICRO-EVOLUTION AND MACRO-EVOLUTION

Darwin called his book 'The Origin of <u>Species</u>'. He sought to explain how new <u>species</u> are formed, how species are descended from preceding species (phylogeny). He thought that the same process which adapts populations to their habitat – natural selection – will eventually create new species. The necessary conditions are variation within populations, some differences among habitats, and sufficiently long time.

Not everybody accepted this idea, even among evolutionists. Goldschmidt (1940), for example, argued that entirely different processes must be at work on the two levels. Natural selection certainly brings about adaptation to local environments – a process he called micro-evolution – but to create new species, genera or families (macro-evolution) complete reorganization of the genetic material, macro-mutations, must occur.

Goldschmidt's idea was rejected by his contemporary biologists, but the two terms – micro- and macro-evolution – are still being used.

Like Goldschmidt, Carson (1982) thought that the formation of new species requires a genetic revolution, an irreversible reorganization of the gene pool. Carson suggested that the genetic system in each individual is in a state of balance, and interacts intimately with the internal physiological functioning of the body as well as with the external environment. To create a new species, the old balance must be disrupted, and a new balance set up (a process of disorganization and reorganization) (see next chapter).

In the evolutionary literature, from Darwin to date, the term species has represented different concepts, leading to disagreement on the interpretation of the facts.

VARIATION, CLASSIFICATION AND SPECIES DEFINITION

Historically, people started from observed variation, classified the organisms they saw into categories by their similarity to each other (= species), and then set rules as to how the membership in these classes should be decided – resulting in a theory of classification. Most of these theories before Darwin did not interpret similarity as an indication of community of descent.

When organisms from a limited geographic region, such as a locality or a country, are assembled (e.g., to create a catalog of the local fauna or flora), it is usually not difficult to assign individuals to species categories. In some cases, species are distinct enough to be recognized by everybody. The lion, leopard and cheetah are easily recognized species belonging to the same family. There are no intermediate forms between them (although hybrids like the 'liger', a cross of the lion and tiger, have been bred in zoos). But in many other cases, the specific differences are not that clear. The problem became even more difficult when domestication and hybridization of plants and animals created intermediate forms.

Species delimitation becomes difficult when collections of organisms from wide geographical areas, or the entire world, are stored in museums and need to be classified. The systematic work of classifying the biological world began with Linneaus and his students in Sweden in the 18th century, and continues to date. As more and more material is amassed, and when the organisms to be classified are small like insects and mites, the borders between distinct groups may become progressively blurred, and the work of the taxonomist progressively harder.

When the investigated organisms are small and unfamiliar, often minute characters are used for species identification and only expert taxonomists can tell them apart. There is sometimes disagreement about which characters should be used as criteria. Larvae and adult organisms belonging to the same species have in some cases been assigned to different species and genera and given distinctive scientific names, while in other cases organisms that look very different have been pooled into a single species. With the incorporation, in the mid-20th century, of electrophoresis – and more recently, DNA sequence comparisons – in taxonomic work, great disparities were found between molecular and previous morphological classifications. Some strikingly different organisms belonging to different species were found to be identical electrophoretically (e.g., Snyder, 1974; Sene and Carson, 1977; Schnell et al., 1978) – while morphologically similar populations of the same species, were found to differ greatly in electrophoretic markers (e.g., Nixon and Taylor, 1978). The issue of which sort of data should be used is still controversial. Among humans, Europeans, Africans, Chinese and aboriginal Australians are different from each other in many characters, yet are unanimously considered a single species, *Homo sapiens*. It all depends on the definition of a species.

Whatever sort of data is used, the important question is, how different two organisms must be in order to be recorded as two different species – or, stated differently, how different may two organisms be and still be considered members of the same species?

DIFFERENT DEFINITIONS OF 'SPECIES'

Some 18th and 19th century philosophers attached different meanings to the term species (see a review in Grant, 1994). The main criterion for setting 'species' apart from 'variety' was the ability to form fertile interspecies hybrids. If two forms did not mate (or their hybrids were sterile) then they belonged to different species – otherwise they were varieties of the same species. This criterion aligns well with the belief in Creation:

> It appears that species have a real existence in nature, and that each were endowed, at the time of their creation, with the attributes and organization by which it is now distinguished... The intermixture of distinct species is guarded against by the aversion of the individuals composing them to sexual union, or by the sterility of the hybrid mule (Lyell, 1853, p. 611).

Linnaeus' species

The foundation of the classification of the biological world was laid down by the Swedish biologist Carlos Linnaeus (Linne'), in his book 'Systema Naturae' (first edition 1735, but the most often cited is the 10th edition of 1758). In the background of this book was the belief that the Creator planned his world in an organized way, and that the classification revealed this organized system. Each unit (species) in this system has a fixed and unchangeable form which can be identified morphologically.

Every new specimen, first of its kind to be encountered, was labeled 'type' – new acquisitions could then be compared to the 'types' to decide if they belong to a known species or that a new species should be named. Following the description of many species, brought in from all the continents, large museums of natural history were built in the 19th century in Europe and in the USA to house the collections. Taxonomic work continues in these institutions to this day.

The breeding and cultivation of plants and animals resulted in individuals with intermediate morphologies, and revealed variation within species. A distinction was made between 'species' and 'varieties'. Forms that could be crossed to produce viable and fertile offspring were considered varieties of the same species. A.R. Wallace criticized the use of this criterion as 'the old theory':

> This [reproductive barrier] used to be considered a fixed law of nature, constituting the absolute test and criterion for a species as distinct from a variety…This law could have no exceptions; because if any two species had been found to be fertile when crossed, and their offspring to be also fertile, this fact would have been held to prove them to be not species, but varieties. On the other hand, if two varieties had been found to be infertile, or their mongrel offspring to be sterile, then it would have been said: these are not varieties, but true species. Thus the old theory led to inevitable reasoning in a circle (Wallace, 1889, p.153).

Darwin's definition

It may come as a surprise at first, to realize that Darwin did not accept 'species' as a fixed entity. He recognized that individuals varied in characters within each species, and that even the offspring of the same pair of parents could be different from one another. With selection working on this variation, the species form cannot remain constant:

> I look at the term species as one arbitrarily given, for the sake of convenience, to a group of individuals resembling each other (Darwin, 1872, p. 39)…
> No one definition has satisfied all naturalists. Yet every naturalist knows vaguely what he means when he speaks of a species (ibid. p. 30).

Darwin's approach is not shared by taxonomists. Although most taxonomists accept the theory of evolution, they continue to define species as fixed units, using morphological criteria – if only for reasons of convenience. As Darwin recognized, minor characters which are of no use to the individuals are preferred as diagnostic characters precisely because they are <u>not</u> likely to be affected by natural selection, and therefore may reflect similarity by descent.

THE 'BIOLOGICAL SPECIES' CONCEPT

With the development of genetics in the 20th century, attempts were made to replace the morphological definition of species with genetic criteria. Statements like "a species comprises organisms that share the same genetic composition" were not useful for lack of means to test individuals for their genetic content. Just before the beginning of the genetic era, Weismann (1886, I: 279, 283) argued convincingly that individuals in a population of sexually-reproducing organisms must be different from one another in hereditary content.

In the 1940s, T. Dobzhansky and particularly E. Mayr worked out a **definition of biological species**, which was accepted in all taxonomic and evolutionary work. Dobzhansky recognized that species are

Closed breeding communities, systems of populations having access to a common gene pool, but reproductively isolated from other populations with separate gene pools (Dobzhansky, 1962).

Mayr (1942) formulated the standard definition now accepted by most biologists:

[Biological species are] groups of populations, which are actually or potentially interbreeding, and are reproductively isolated from other such groups (Mayr, 1942).

The critical criterion for species by this definition is thus the potential for free gene flow within the species, and a reproductive barrier preventing gene flow with other species [ironically, the reliance on reproductive barriers is a regression to the 'old system' which was criticized as circular reasoning by Wallace!].

Later taxonomists insisted, further, that hybrid sterility – or otherwise - should be tested in nature (interspecies hybrids obtained in captivity do not count as evidence for gene flow).

<u>Barriers to gene flow</u> exist in nature in many forms. Geographical barriers that restrict the dispersal of individuals – like oceans, rivers, deserts or mountain ranges – are obviously also barriers to interbreeding. Ecological differences in soil, flowering times, or temperature tolerance serve the same purpose. In animals, differences in courting rituals (e.g., in *Drosophila*, fishes, and birds) form decisive barriers to mating between species. Last come differences in copulatory organs (in insects) and

physiological and genetical barriers, preventing fertilization or resulting in sterile hybrids (Mayr, 1963; Dobzhansky, 1970).

Following Darwin, Mayr recognized that from an evolutionary standpoint, a species is not an absolute, fixed unit: it is a dynamic entity. There exists a gradient of stages of reproductive barriers. Some species are different morphologically, but can still hybridize (**'incipient' species**). Others are morphologically similar, but reproductively completely isolated (**'sibling' species**). Intermediate stages between the two also occur. These are different stages of the process of speciation (see next chapter). Examples of all these stages were described in *Drosophila*.

The biological species concept is widely accepted today. Taxonomists, who classify organisms exclusively by morphological characters, believe that their decisions are sound because morphological differences between their species would not have been maintained unless a reproductive barrier existed between them.

OTHER DEFINITIONS

It may not be an exaggeration if I say that there are probably as many species concepts as there are thinking systematists and students of speciation (Mayr, 1942, p. 115).

A reproductive barrier can be a decisive step in the evolution of new species, because it transforms an 'open' genetic system – which can absorb new material – to a 'closed' system, which can evolve independently (Dobzhansky, 1970). But Dobzhansky recognized that taxonomists cannot usually know whether a reproductive barrier ever existed – most of the material they examine is dead. Moreover, the biological species definition is not applicable to asexually-reproducing organisms, particularly to bacteria and viruses which reproduce asexually during most of their life. These organisms are of great importance to humans, and their identification and classification has a great practical value. Dobzhansky suggested that the difficulty may be solved by attaching different meanings to the term 'species' in evolutionary and in taxonomic studies: in the latter use, 'species' is only a descriptive term with no genetic implications (Dobzhansky, 1970).

The British biologist, Julian Huxley (grandson of Darwin's friend and ally, Thomas Henry Huxley) suggested that the term species could be flexibly defined and qualified to provide different definitions for different contexts. 'Geographical' species may show similar adaptations to climatic conditions, but may not necessarily show behavioral or genetical reproductive barriers. 'Ecological' species may show strong habitat specializations (ecotypes), sometimes together with ecological reproductive barriers. 'Genetical' species may show behavioral, mating, or hybrid sterility barriers, even if they are hard to tell apart morphologically (Huxley, 1942).

One objection to the 'biological species' definition is that it relies on negative evidence (absence of gene flow) and not on positive characteristics of the species. Paterson (1985) accordingly suggested that characters that enable organisms to recognize their species mates should be used as criteria (the 'recognition species' concept): Pre-mating isolating mechanisms can form when the recognition signals that attract the sexes are altered, even without geographical or other barriers. Recognition criteria require detailed observations on the mating behavior of individuals, which are not generally available outside of laboratory populations, and Paterson's suggestion has not been widely accepted by biologists.

Difficulties in Practical Application of the Biological Species Concept

In a critical study, Sokal and Crovello (1970) showed that the biological species definition is not applicable to any practical situation likely to be encountered in nature, and cannot be used to determine species boundaries even in sexually-reproducing organisms. In their hypothetical example they tried to decide how many species are present in a given geographical area. Going step by step through a flow chart of the process, they show that at every step, the use of the biological species definition is impractical, because the crucial information on mating ability and reproductive success is unobtainable, and that morphological similarity must invariably be used instead (it is impossible to test all possible combinations of males and females of different organisms to find out whether they will mate at all, and if so whether their hybrids are fertile – without relying on morphological similarity to group individuals together).

Sokal and Crovello advocate the use of purely morphological criteria – as is done anyway by taxonomists – without the need to assume reproductive barriers: the biological species concept is not really necessary. They suggest that the term 'species' is useful and necessary in classifying the biological world. But they insist it should be defined quantitatively based on many phenetic characters (see below).

MULTIVARIATE MORPHOLOGICAL DEFINITIONS: NUMERICAL TAXONOMY

Relying on a few diagnostic characters for species identification may lead to serious biological errors. For example sexual dimorphism may place males and females of the same species – in different taxa. Larval and adult chironomid midges were classified in different families. Many gall-inducing aphids were given different generic and species names when they were collected on different host plants and in different countries, resulting in many synonyms.

Species should be characterized by the similarity in as many characters as possible–not by the presence or absence of a few characters. The simultaneous comparison of many characters requires the use of computers for calculating similarity and revealing taxonomic structure (similarity is not an all-or-none character: there is a gradient between 'totally different' and 'identical'). Several statistics were offered to measure similarity (Sokal and Sneath, 1963; Sneath and Sokal, 1973).

Sokal and Sneath suggested a computer-based multivariate approach, called Numerical Taxonomy (NT). The end product of NT is a hierarchical diagram referred to as a phenogram, reflecting the phenotypic similarity of the Operational Taxonomic Units (OTUs) – which may be individuals, species, or any other units – to each other (Fig. 21.1). Once the similarity phenogram is determined, taxonomists may decide what level of similarity should delimit a species. Disagreement on this point will not affect the hierarchy itself.

Many hundreds – perhaps thousands – of papers were published in the 1970-1980s using NT and testing its applicability from different angles. One advantage of NT is that the similarity of organisms can be assessed from sets of data collected by non-professional workers – even children – and even determined by machines (Sokal and Rohlf, 1970, 1980).

There may be a revival of this approach in the future: the current scientific and public interest in biodiversity has created a great need for specialist taxonomists – but their number in the world is very small ("although much biological research depends upon species diagnoses, taxonomic expertise is collapsing". Hebert et al., 2003). In many recent studies, non-professionals are classifying the organisms into 'morphospecies', recognizable morphologically. Practical demand may force ecologists to rely more on computers (and NT), especially now that efficient handling of computers is much more widespread among scientists than it was 20 years ago.

The multivariate approach to classification developed an important branch, dealing with the study of phylogenetic relationships. This method, called **Cladistics**, emphasized that similarity among organisms reflects their community of descent. This subject is dealt with in Chapter 23.

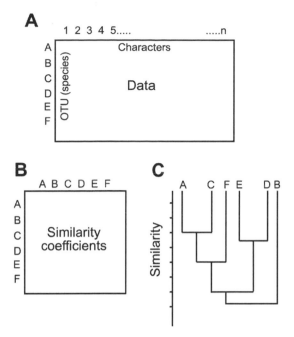

Fig. 21.1 Stages in the preparation of a phenogram by the NT methodology. (A) A matrix of raw data, with measurements of n characters on each OTU (species A – F). (B) A matrix of similarity coefficients between pairs of OTUs. See Sokal and Sneath for measures of similarity. (C) The resulting phenogram after clustering the OTUs by their similarity to each other.

GENETIC IDENTITY AND GENETIC DISTANCE

What is the minimal amount of genetic similarity that justifies the inclusion of populations in the same species? If a suitable measure can be agreed upon, genetic similarity could perhaps be a quantitative way of defining species.

The most often used indices of genetic similarity are Nei's coefficients of genetic identity, **I**, and genetic distance, **D** (Nei, 1972) (Box 21.1)

Box 21.1 **Calculations of Nei's genetic identity and distance**

If x_i and y_i are the frequencies of allele i in two populations X and Y, and n is the number of alleles observed, define

$$J_x = \Sigma^n x_i^2/n \quad J_y = \Sigma^n y_i^2/n \quad J_{xy} = \Sigma^n x_i y_i/n$$

then the genetic identity is $\quad I = J_{xy}/\sqrt{(J_x J_y)}$

and the genetic distance is $\quad D = -\ln I$

A comprehensive study of electrophoretic variation was carried out in *Drosophila willistoni* in an attempt to answer the question posed above (Ayala, 1975). In this group of 15 species, there are six morphologically indistinguishable sibling species. In some of the other species there are subdivisions (subspecies) with incomplete reproductive isolation. Thirty-six electrophoretic loci were examined. The levels of **I** and **D** are listed below (Table 21.1).

Table 21.1 Mean genetic identity (**I**) and distance (**D**) in *D. willistoni* (from Ayala, 1975)

	I	D
Among populations within species	0.972	0.028
Among subspecies within species	0.795	0.214
Among sibling species	0.583	0.581
Among 'good' species	0.352	1.056

These data give the impression that there is some relationship between genetic distance and the level of reproductive isolation in this closely-related group of species: the more reproductively isolated the species, the greater the genetic distance among them. However, values of genetic distance and identity vary between wide limits, even within *Drosophila* species (as can be seen from the figures in Ayala, 1975), and even more so between different organisms (Bruce and Ayala, 1979). Thus the values in Table 21.1 cannot be used as quantitative criteria for species status.

Further, there is <u>no proof</u> that reproductive isolation is a consequence of the accumulation of genetic differences! In one study, the genetic similarity among populations of the snail, *Cepaea nemoralis*, from Italy, England and the USA was very low (I = 0.53, based on 20 electrophoretic markers) but crosses between snails from different countries were fertile, despite the wide geographical distances between them and the low vagility of snails (Johnson et al., 1984).

All 103 species of Hawaiian picture-winged *Drosophila* are different morphologically, and most are also reproductively isolated from each other by geographical and ecological barriers, but their chromosomal inversions and banding patterns are very similar. Cytological comparisons between populations of picture-winged Hawaiian *Drosophila* in different stages of reproductive isolation, detected no consistent differences in inversion frequencies, although hybrid sterility and other reproductive isolation phenomena are often associated with inversions (Carson, 1981a, b). There was no relationship between the chromosomal differences between species and the age of the islands on which they are endemic. Two species, endemic to the 'big island' of Hawaii – *D. heteroneura* and *D. silvestris* – which are very different morphologically and behaviorally, are identical in chromosomal pattern (Carson, 1981c).

MOLECULAR IDENTIFICATION OF SPECIES?

The advances in DNA sequencing technology on the one hand, and the shortage of specialist taxonomists for the sorting of the huge number of insects in biological conservation studies on the other, suggest to some that DNA technology could be employed in species determination. Hebert et al. (2003) argue that the solution to the species identification problem is the construction of DNA sequences as species identification markers. These authors recommend that segments of the mitochondrial cytochrome C oxidase I gene (COI) can be used as a universal species 'barcode'. The authors argue that even a sequence as short as 15 BP can provide more than enough individual markers to cover all life if each taxon was uniquely banded. (These barcodes can be read by machines, similar to product codes in supermarkets.)

Hajibabaei et al. (2006) applied this approach in a study of 521 species in three families of tropical Lepidoptera in Costa Rica. The COI barcodes enabled the correct identification of 97.9% of the species recognized by previous taxonomic work.

However, species identification by barcodes (= classification, assignment of individuals to known species) is based on a previously determined taxonomic structure, and is not the same as taxonomy (= delimitation of species boundaries and taxonomic structure). The use of DNA barcodes will require a new species definition.

Rubinoff and Sperling (2004) admit that mtDNA is "one of the best molecular markers for evaluating species boundaries for conservation and diversity studies", but argue that decisions based on mtDNA should be calibrated with morphological, ecological and behavioral data.

Will and Rubinoff (2004) reject outright the idea of a universal species barcode and the idea of replacing ecological and morphological species definitions by a DNA barcode. There is no simple formula for deciding the length of the sequence required for a barcode and there is no standard by which to calibrate this decision – even between families of the same order. Will and Rubinoff conclude that molecular data are an important and powerful part of taxonomy and systematics, and have an indisputable role in the analysis of biodiversity, but cannot be a substitute for studying whole organisms.

"The notion that there is an inherent supremacy of DNA data versus other types of character data for all taxonomic questions and circumstances is wrongheaded" (Will and Rubinoff, 2004, P. 54).

Literature Cited

Ayala, F.J. 1975. Genetic differentiation during the speciation process. Evolutionary Biology 8: 1-78.

Bruce, E.J. and F.J. Ayala. 1979. Evolutionary relationships between man and the apes: electrophoretic evidence. Evolution 33: 1040-1056.

Carson, H.L. 1981a. Chromosomes and evolution in some relatives of *Drosophila grimshawi* from Hawaii. pp.195-205. In: R.L. Blackman, G.M. Hewitt and M. Ashburner (eds.), Insect Cytogenetics, Blackwell, Oxford, U.K.

Carson, H.L. 1981b. Chromosomal tracing of evolution in a phylad of species related to *Drosophila hawaiiensis*. pp. 286-297. In: M.J.D. White, W.R. Atchley and D.S.Woodruff (eds.), Evolution and Speciation, Cambridge University Press, Cambridge, UK.

Carson, H.L. 1981c. Homosequential species of Hawaiian *Drosophila*. Chromosomes Today 7: 150-154.

Carson, H.L. 1982. Speciation as a major reorganization of polygeic balance. pp. 441-443. In: Mechanisms of Speciation, A.L. Liss Inc, New York, USA.

Darwin, C. 1872. The Origin of Species. 6th ed. Murray, London, UK.

Dobzhansky, T. 1962. Mankind Evolving. Yale University Press, Boston, USA.

Dobzhansky , T. 1970. Genetics and the Evolutionary Process. Columbia University Press, New York, USA.

Goldschmidt, R. 1940. The Material Basis of Evolution. Yale University Press, Boston, USA.

Grant, V. 1994. Evolution of the species concept. Biologisches Zentralblatt 113: 401-415.

Hajibabaei, M., D.H. Janzen, J.M. Burns, W. Hallwachs and P.D.N. Hebert. 2006. DNA barcodes distinguish species of tropical Lepidoptera. Proceedings of the National Academy of Sciences, USA, 103: 968-971.

Hebert, P.D.N., A. Cywinska, S.L. Bell and J.R. de Waard. 2003. Biological identifications through DNA barcodes. Proceedings of the Royal Society, London B 270: 313-321.

Huxley, J. 1942. Evolution: The Modern Synthesis. George Allen & Unwin, London, UK.

Johnson, M.S., O.C. Stine and J. Murray. 1984. Reproductive compatibility despite large scale genetic divergence in *Cepaea nemoralis*. Heredity 53: 655-665.

Lyell, C. 1853. The Principles of Geology. 9th ed. Murray, London, UK.

Mayr, E. 1942. Systematics and the Origin of Species. Columbia University Press, New York, USA.

Mayr, E. 1963. Animal Species and Evolution. Harvard University Press, Boston, USA.

Nei, M. 1972. Genetic distance between populations. American Naturalist 106: 283-292.

Nixon, S.E. and R.J. Taylor. 1978. Large genetic distances associated with little morphological variation in *Polycelis coronata* and *Dugesia tigrina* (Planaria). Systematic Zoology 26: 152-164.

Paterson, H.E.H. 1985. The recognition concept of species. pp. 21-29. In: E.S. Vrba (ed.), Species and Speciation, Transvaal Museum, monograph #4, Pretoria.

Schnell, G.D., T.L. Best and M.L. Kennedy. 1978. Interspecific morphologic variation in kangaroo rats (*Dipodomys*): degree of concordance with genic variation. Systematic Zoology 27: 34-48.

Sene, F.M. and H.L. Carson. 1977. Genetic variation in Hawaiian *Drosophila*. IV. Allozymic similarity between *D. silvestris* and *D. heteroneura* from the island of Hawaii. Genetics 86: 187-198.

Sneath, P.H.A. and R.R. Sokal. 1973. Numerical Taxonomy. Freeman, New York, USA.

Snyder, T.P. 1974. Lack of allozymic variability in three bee species. Evolution 28: 687-688.

Sokal, R.R. and P.H.A. Sneath. 1963. Numerical Taxonomy. Freeman, New York, USA.

Sokal, R.R. and T.J. Crovello. 1970. The biological species concept: a critical evaluation. American Naturalist 104: 127-153.

Sokal, R.R. and F.J. Rohlf. 1970. The Intelligent Ignoramus: an experiment in numerical taxonomy. Taxon 19: 305-319.

Sokal, R.R. and F.J. Rohlf. 1980. An experiment in taxonomic judgement. Systematic Botany 5: 341-365.

Rubinoff, D. and F.H.A. Sperling. 2004. Mitochondrial DNA sequence, morphology and ecology yield contrasting conservation implications for two threatened buckmoths (*Hemileuca*; Saturnidae). Biological Conservation 118: 341-351.

Wallace, A.R. 1889. Darwinism. Murray, London, U.K.

Weismann, A. 1886. Essays on Heredity. Oxford University Press, Oxford, UK.

Will, K.W. and D. Rubinoff, 2004. Myth of the molecule: DNA barcodes for species cannot replace morphology and classification. Cladistics 20: 47-55.

CHAPTER 22

Formation of New Species (Speciation)

What Lamarck then foretold had come to pass: the more new forms have been multiplied, the less are we able to decide what we mean by a variety, and what by a species. In fact, zoologists and botanists are not only more at a loss then ever how to define a species, but even to determine whether it has any real existence in Nature, or is a mere abstraction of human intellect (Lyell, 1973, p. 435).

HISTORICAL REVIEW

When Linneus published his Systema Naturae in the 18th century, no one doubted that the characteristics of each species were fixed and unchangeable since their creation. There was no need to explain how new species were formed: no new species were added to the world – all of them were present from the start, although not all are known to humans, and zoologists and botanists discover new species periodically. This belief was still prevalent in the first half of the 19th century. The geographic distribution of animals in diverse habitats was explained by the biblical story of Noah's Ark: when the ark landed on Mount Ararat (in Armenia) and the water receded, animals spread from there to all parts of the world. Species with diverse habitat requirements could have found a temporary refuge on that mountain, since its summit has a permanent ice cap suitable for arctic animals, and its base is in a hot and dry land, with temperate areas in between (Haeckel, 1876).

Charles Lyell believed in the fixity of species, but not in the dispersal from a single point. His interpretation of a gradual, 'special creation' of species at different places and times suggests a modern ecological outlook (Wool, 2001): there must have been a long interval after the creation of grasses, before the first pair of herbivores could be allowed to multiply, otherwise they would have consumed all the food and died of starvation. Similarly, herbivores must have been allowed a long time for reproduction before the first carnivores were created. This of course required the indefinite extension of the biblical 6-day limit of creation.

Darwin took with him the first edition of Lyell's book when he sailed on board the Beagle. He was enthusiastic about the book. But his attitude towards species and Creation was very different from Lyell's. Darwin thought that species can and do vary, and that they change with geological time. New species developed by the accumulation of small, favorable variations by natural selection. The idea was expressed by Dobzhansky (1962) and accepted by many other Darwinians:

> Species arise gradually by the accumulation of gene differences, ultimately by summation of many mutational steps which may have taken place at different times. And species arise not as single individuals, but as diverging populations (Dobzhansky, 1962, p. 181).

In the early 20th century, following the the discovery of Mendel's paper in 1900 and the establishment of genetics as a science, some geneticists (the 'mutationists') argued than new species arise suddenly as a direct result of mutations (the last sentence in Dobzhansky's quotation above is a reply to these arguments). The Dutch botanist, H. de Vries, one of the discoverers of Mendel's paper, wrote that natural selection "explains the survival of the fittest, but cannot explain the arrival of the fittest":

> Striking departure from the previous appearance of a species...each mutation sharply and completely separates the form, as an independent species, from the species from which it arise (cited by Mayr, 1942, p. 65-66).

It is interesting that the majority of mutations in plants, described by de Vries, were cases of polyploidy – which in fact causes an immediate reproductive barrier! Most species of polyploids reproduce asexually, and are infertile when crossed with other non-polyploids. A large number of existing plant species may have evolved by this mechanism (Ayala, 1975).

The biologist St. George Mivart, one of Darwin's critics, questioned whether species could be formed gradually by natural selection. What could be the selective advantage of organs – useful as they may become in their final form, like wings in birds – when they first appear as mere rudiments? The first stubs of feathers could not keep the animal warm, and rudimentary wings could not be used for flight. The transition from aquatic to terrestrial life (amphibian to reptile, for example) required the enclosure of yolk in a hard shell, a structure for piercing the egg shell in the embryo, structures for disposal of metabolic wastes etc. All these changes should have occurred simultaneously, because each change separately was not sufficient for life to begin on land (Mivart, 1871). Darwin did not find convincing replies to these objections.

HOW DO NEW SPECIES EVOLVE?

Julian Huxley (1942) described the processes of speciation as a network of trial and error steps in different directions, most of which are blind alleys. The evolution of a line may become a dead-end alley because of excessive specialization – for example, strict preference for some kind of food, or specialized anatomical structures, which may become useless if conditions change. One possible example of such a dead end may be the giant pandas in China, feeding on the leaves of special bamboos: they are now threatened with extinction because the range of suitable vegetation in their natural habitat has severely decreased.

Opinions on speciation are affected by the concept of species that the writers have in mind. If 'species' is just a convenient term to describe a group of similar individuals, as Darwin suggested, then new species may evolve by a gradual accumulation of small, heritable morphological changes. But most biologists see the species as closed genetic entities (the biological species concept), and speciation as a process of formation of reproductive barriers.

Accordingly, natural populations in different levels of reproductive isolation – subspecies, sibling species and incipient species, as described above for *Drosophila* – are regarded as stages in the speciation process. Reproductive barriers were discovered between groups with no electrophoretic, chromosomal or morphological differences between them: speciation may be unrelated to the level of genetic variation in the groups (Templeton, 1981).

GEOGRAPHIC (ALLOPATRIC) SPECIATION

A new species develops if a population which has become geographically isolated from its parental species acquires during its period of isolation characters which promote or guarantee reproductive isolation when the external barrier breaks down (Mayr, 1942, p. 155).

Groups separated by geographical isolation are originally species only in posse. Their separation into good species is a subsequent process, accompanying the process of character divergence (Huxley, 1942, p. 383).

The most common scenario for speciation begins with the separation of part of a previously-continuous population by a geographic barrier (Mayr, 1963; Bush, 1975; Ayala, 1975). The formation of a geographic barrier reduces or prevents gene flow between the separated subpopulations, which are then free to accumulate different mutations – making them progressively different from each other. Inbreeding and natural selection drive them farther apart, while the geographic barrier prevents mixing of the gene pools. If the barrier persists long enough, the subpopulations will become reproductively isolated from each other, and will not mix even if the barrier is removed: they have evolved into two separate species (Fig. 22.1A).

The attainment of a 'good species' status requires the formation of reproductive isolation. This process is likely to take place first at the boundaries between the populations when they meet again after a period of geographic separation (Bush, 1975). Such a 'secondary' isolating mechanism may be a behavioral recognition barrier, for example in species which requires a male courtship display to ensure successful mating: the courtship patterns may diverge to the extent that females may not accept the courting males. Behavioral barriers to fertilization are referred to as 'pre-zygotic' mechanisms, and are considered the cheapest (evolutionarily speaking) because they do not involve loss of zygotes.

Mutations altering courtship and mating success were described in male *Drosophila*. Although some interspecies matings do take place between two closely-related species, *D. persimilis* and *D. pseudoobscura*, the males of each species, when given a choice, tended to mate preferentially with females of their own species (*D. pseudoobscura* males: 84.3% versus 7%; *D. persimilis* males: 79% versus 22.5%. Dobzhansky, 1955).

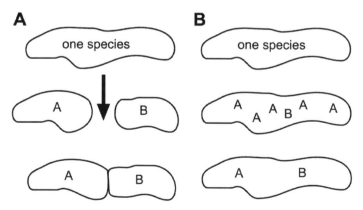

Fig. 22.1 Different scenarios for speciation. A. Allopatric speciation by a geographical barrier. B. Sympatric speciation by adaptation to host plants A and B.

Behavioral isolation between two sibling species of *Drosophila* was described in detail by Coyne (1993). *D. simulans* (SIM) is distributed world-wide – especially in human habitats. *D. mauretanica* (MAU) is endemic to Madagascar. In interspecies laboratory crosses, F1 males are sterile. The F1 females are fertile, and backcrosses to males of both strains produce fertile offspring. But the <u>mating duration</u> of male MAU × female SIM is particularly short (5-8 min.) compared with SIM × SIM or MAU × MAU matings, (averages 25 and 15 min., respectively). Female SIM endeavor to kick off the males of the wrong species, and thus interrupt sperm transfer. The numbers of offspring in inter-species matings are much lower than normal. Genes affecting this behavior were located on at least two chromosomes (Coyne, 1993).

Male flour beetles, *Tribolium confusum* (CF), do copulate with females of *T. castaneum* (the mating is sterile since there is no fertilization), but males of *T. castaneum* (CS) do not mate with *T. confusum* females. This is outstanding in view of the fact that *Tribolium* males are indiscriminant and often mate homosexually (Graur and Wool, 1982).

Populations may speciate geographically along environmental gradients. Geographical – ecological speciation may begin in the formation of **ecotypes**, in particular in plants. The plant *Nigella arvensis* inhabits different habitats in Israel, from the Mediterranean shoreline to a few kilometers inland. The coastal individuals are prostrate while the inland plants grow upright. There are differences also in leaf shape and other characters. Some of the differences are genetic, but gene flow across the range is possible and ecotype crosses in the laboratory are fertile. The coastal ecotype is an adaptation to the salt spray blown from the sea with the west winds (Waisel, 1959).

A secondary isolating mechanism may result if a mutation caused a change in the morphology and/or anatomy of the sexual organs of either the male or the female, which prevents normal sperm transfer and fertilization. In insects, sexual organs vary greatly among species – for example in the Diptera, where the shape of the male copulatory organ provides major taxonomic characters for separating species. In plants, interspecies hybridization is prevented by the inability of pollen grains to germinate on another species' stigmas.

'<u>Post-zygotic</u>' mechanisms such as hybrid sterility or mortality, which preclude gene flow between species, are more 'wasteful'. Hybrid sterility was reported among species of *Drosophila*. Intermediate stages are also known. For example, male F1 hybrids of *D. persimilis* and *D. pseudoobscura* are sterile, but female F1 may be back-crossed to either male and produce viable – albeit weak – offspring.

SYMPATRIC SPECIATION

The question of whether new species can arise without a geographic barrier (i.e, <u>in sympatry</u>) has long been controversial. According to the definition of the Biological Species, sympatric speciation requires that a reproductive barrier be formed <u>within</u> a population (Fig. 22.1 B). If the barrier is a result of mutation, it should be a rare event, and the chance of the same mutation in two individuals of opposite sex is remote. Even if the resulting reproductive barrier is not complete, the mutant will still have to mate with a non-mutant to leave any progeny (see Chapter 5). It is considered unlikely that a viable population of the mutant will be established sympatrically in this way.

Nevertheless, situations exist in nature where sympatric speciation is the favored explanation, and scenarios for its evolution have been proposed. For a recent review of models of sympatric speciation, see Fry (2003).

One scenario suggests that mutations causing translocation in non-homologous chromosomes may lead to speciation in sympatry. Such translocations cause up to 75% sterility of hybrids. A gamete with a translocated chromosome uniting with a normal gamete, often result in a non-viable zygote because of imbalance of the genetic material. But the union of two mutant, translocated gametes is often viable (Wagoner et al., 1969). Suppose that a chromosome is translocated in a sexual stem cell in one individual. When united with normal gametes, some of the gametes produced by this individual may succeed in producing viable heterozygotes. These may give rise, by recombination, to viable males and females homozygous for the translocation, which will have a full genetic complement and breed together, sympatrically with, but genetically isolated from the parent species.

The evolution of two closely-related species of flour beetles (*Tribolium*) may be explained by this scenario. *T. castaneum* has nine pairs of autosomes plus an x, y pair. *T. confusum* has eight autosome pairs and an x, y pair, but the x is larger than in the congeneric species. Smith (1952) suggested that one of the autosomes in *T. castaneum* ancestor was translocated to the x, leaving the homologue of the autosome as a new y chromosome. Some genetic data in support of this scenario have been published.

HOST RACES IN FRUIT FLIES

Bush (1975) argued that sympatric speciation is a viable possibility, and may occur by differential adaptation to different host plants. Host races may be a step before complete isolation of species (a **host race** is defined as "a population of a species, living on and showing a preference for a host which is different from the host or hosts of other populations of the same species"; Bush and Diehl, 1982). Huettel and Bush (1972) described a scenario which is applicable to a speciation event in fruit flies (Tephritidae). The presence of two host plants within the same geographic site leads to ecological separation of the fly population into races. The two host races may then evolve to become good species, provided that 1) host selection is a heritable character, and 2) mating is more frequent between flies from the same host plant than between flies from different hosts.

Bush and Diehl (1982) considered a model which requires two genes: a survival gene and a host-recognition gene. A mutation enabling survival on a previously-unusable host will make the population polymorphic, but will not, by itself, make the mutant a host-race. Similarly, a mutation allowing recognition of the new host as suitable, will not by itself lead to speciation. But suitable mutations in both genes will enable the mutant to recognize the new host as well as to develop and survive on it, which other populations are unable to do. This situation is preparatory to speciation. Also, since individuals are assumed to mate on the host, the difference in host choice itself brings about partial reproductive isolation. "An association (linkage disequilibrium) between an allele at a viability locus, which confers superior fitness on one of the two hosts, with an allele at a second locus specifying a preference for the same host, will be directly favored" (Bush and Diehl, 1982).

Strong support for this scenario of sympatric speciation came from the work of Feder et al. (1988). The fruit fly *Rhagoletis pomonella* (Tephritidae) is endemic to North America and its original host was hawthorn. The female fly oviposits into the fruit where the larvae develop. Mature larvae drop to the ground to pupate (with or without the fruit) and emerge as adults the next year. When apples were introduced commercially into the USA in the early 19th century, they became a new favorable host for the flies. Apples and hawthorn are often located quite close together at the same sites in North America.

Feder et al. (1988) found consistent differences in allele frequencies at six polymorphic electrophoretic markers between *R. pomonella* caught (or collected as larvae) on apple and hawthorn trees in Michigan. Despite the absence of geographical barriers to gene flow, the populations feeding on the two hosts have not mixed. This was confirmed in consecutive years and in samples from several research sites in Michigan (Feder et al., 1988). A geographical cline of decreasing frequencies was observed at sites south of Michigan, but the flies on the two hosts remained different (Feder and Bush, 1989). The flies mate on the fruit, and the adult flies do not fly far from their natal tree – increasing the probability of mating of flies from the same host. Other behaviors that may keep the races separate were also reported. 'Apple' females tended to be preferentially attracted to apple fruit for oviposition, and apples were accepted much more often by ovipositing females of apple origin than by females of hawthorn origin. The males emerging from the two kinds of fruit showed a similar preference – apple males spent more time on apples than hawthorn males (Prokopy et al., 1988). There is also temporal separation in the phenology of the two races: the apple fruit matures a few weeks earlier than hawthorn. Apple flies enter diapause early – giving the hawthorn flies an advantage (Feder et al., 1997). The flies were better protected from parasites in the larger apple fruit (percentage parasitism in apples was 13%, in hawthorn 46%; Feder, 1995).

A similar example for sympatric differentiation is described by Itami et al. (1997). The goldenrod gall fly, *Eurosta solidaginis* (Tephritidae), parasitizes two species of *Solidago* – *S. altissima* and *S. gigantea*, in northern USA. The females lay eggs on the plant and the maggots dig their way into the stem, where they induce a large, spherical gall. Like in *Rhagoletis*, there is a temporal difference in emergence times of *Eurosta* on the two hosts, and mating tends to occur on the host where the larvae grew. MtDNA and electrophoretic studies showed that the host races of *E. solidaginis* are different, but there was no indication of reproductive isolation.

[NOTE: Although the host-race differences are suggestive, there is no proof that these host races will eventually become 'good' species!!]

BOTTLENECKS: FLUSH-CRASH CYCLES

Based on the extraordinary speciation of Drosophilids in the Hawaiian islands (see below), Carson (1975) suggested that reducing population size to very small numbers (crash), followed by population growth again to large size (flush), if repeated several times, could lead to speciation.

The 'flushed' population becomes large after each catastrophe, but its gene pool, which was necessarily greatly diminished by the bottleneck at the crash stage, remains small. Recombination during the flush stage may give rise to new genotypes. Some of these would have been disadvantageous and selected against if they had occurred before the catastrophe, but may survive in the less restrictive conditions promoting the flush stage (small populations in a rich habitat, low competition). A series of such crash-flush cycles – with the accompanying effects of random drift and founder episodes – may, by this model, result in genetically different populations, leading to reproductive isolation and speciation (Carson and Templeton, 1984).

An accidental event in the laboratory supported Carson's model. A laboratory population of *Drosophila silvestris*, polymorphic for three chromosomal inversions, was almost wiped out when an incubator malfunctioned: almost all larvae were killed, and most of the adults were sterilized by the sudden increase in temperature. No more than three adults of each sex survived and were used to start a new population. After the recovery (the flush stage), all three inversions were still present despite the bottleneck, and several new combinations were found that were not recorded before the crash. The genetic variation <u>increased</u> after the catastrophe (Carson, 1990).

Hawaiian drosophilids

Carson based his model on the biology and chromosomal variation of Hawaiian drosophilids. Hundreds of species of drosophilids exist in the Hawaiian archipelago – more than in any other part of the world – and many of them are endemic to Hawaii. More than 100 of these species belong to a group called 'picture-winged', which Carson studied in detail.

The Hawaiian archipelago consists of a chain of oceanic islands, of volcanic origin, which were formed above a fixed 'hot spot' in the earth's crust on the ocean floor. In the quiescent interval after the volcanic eruption that created each island, the new island moved north-westwards away from the 'hot spot' with the oceanic plate, and was left at the mercy of the waves and other erosion factors. New islands appeared in succession (the most northerly islands have long disappeared under the ocean surface but can be detected by sonar). An examination the potassium/argon ratio in the volcanic rocks indicates that the oldest island, Kauai, is 5.6 million years old. The 'youngest', the Big Island of Hawaii, is 'only' 700,000 years old, and has an active volcano, Kilauea, which erupts periodically.

Each island has its share of endemic picture-winged *Drosophila*. The volcanic origin of the islands and the distance of the archipelago from the American continent precludes frequent colonization of the islands from the mainland. New islands must have been colonized by a few migrants – perhaps single fertilized females – from older islands. Even within the Big Island, frequent lava flows redivide the terrain into smaller units ('Kipukas') where pockets of vegetation remain (on which the fly larvae feed) while the rest is destroyed by the lava. This creates a system of ecological islands within islands. A series of flush-crash cycles is a plausible model for the speciation of these flies. An examination of the chromosomal patterns of the flies enabled Carson to suggest probable pathways for the inter-island migration of the founders of the fly populations.

Experimental evidence

Powell (1978) started a polymorphic *Drosophila* population in a large cage, with plenty of food. When the population size became very large, Powell randomly selected 12 pairs of flies, and let each pair found a new daughter population. The daughter populations were permitted to 'flush' again for a few generations after the bottleneck and reach large sizes, before 'crashing' them to one-pair bottlenecks again. The procedure was repeated four times.

Reproductive isolation was estimated after each generation from mating frequencies between flies from pairs of sister populations, using the statistic $I = (A - B)/(A + B)$, where A is the number of matings of males and females from the same population, and B the number of matings of males of one population with females from the other. After the first bottleneck there was no indication of reproductive isolation, but after the fourth episode, three of the surviving eight populations showed a tendency of preferential mating within populations – a first step toward differentiation.

THE EFFECTS OF GENETIC BOTTLENECKS: LABORATORY STUDIES

A series of experiments with laboratory populations of houseflies, spanning 10 years, were designed to test the flush-crash cycle model. The question asked was, whether a series of bottlenecks would not deplete the genetic variation in the population and thus lead not to speciation, but – on the contrary – to extinction.

The same experimental lines were used in all experiments. From a large, genetically variable base population of houseflies, 50 replicate families each were started with propagule size of one, four, and 16 pairs of flies. These experimental replicates were maintained for a number of generations, and the same bottleneck size enforced at the start of every generation (many lines were lost due to inbreeding depression). After each bottleneck, each replicate was allowed to 'flush' for a few generations until it reached a size of approximately 1000 pairs. A replicated control population (not bottlenecked) was also maintained. Eight morphometric traits were measured in samples from each replicate, and heritability of the traits was calculated from parent-offspring regression (Chapter 10).

In the first study (Bryant et al., 1986a, b) a <u>single</u> bottleneck did cause a reduction in fitness, as expected: viability, individual weight and total biomass per replicate decreased with smaller bottleneck sizes. The proportion of infertile pairs of flies was much higher in the bottlenecked populations than in the control, and was inversely related to bottleneck size as predicted from inbreeding depression: one-pair lines – 33% infertile pairs; four-pair lines – 16%; sixteen-pair lines – 11% ; control: 4% infertile pairs.

In contrast with the measures of fitness, no decrease in the additive genetic variance was observed. On the contrary, the additive genetic variance of morphological traits <u>increased</u> after the crash, and was higher in the bottlenecked populations than in the control. The greatest increase was in the smallest bottleneck. This was an important support for Carson's model:

> If bottlenecks serve to increase additive genetic variation – if only temporarily – the rate of evolution in a new environment will be accelerated (Bryant, 1986a).

The next study examined the effects of a series of three or five crash-flush cycles on electrophoretic variation at four polymorphic loci, totaling 13 alleles, in the same experimental lines (McCommas and Bryant, 1990). No loss of heterozygosity was observed after repeated bottlenecks, but some alleles were lost from the one- and four-pair bottleneck lines. This was the expected result of genetic drift in populations of small bottleneck size (McCommas and Bryant, 1990).

The conclusion from these experiments has important consequences.

> Additive genetic variance for all bottleneck sizes rose above the level of the outbred control in response to the first bottleneck, and remained comparable to it or higher than that of the control over most of the successive bottleneck episodes (Bryant and Meffert, 1992). Bottlenecked populations can retain sufficient genetic variability for fitness, and they rapidly adapt to a new environment (a laboratory one, though) as expected from the founder-flush theory (Bryant and Meffert, 1990).

Literature Cited

Ayala, F.J. 1975. Genetic differentiation during the speciation process. Evolutionary Biology 8: 1-78.

Bryant, E.H., S.A. McCommas and L.M. Combs. 1986a. The effects of an experimental bottleneck upon quantitative genetic variation in the housefly. Genetics 141: 1191-1211.

Bryant, E.H., L.M. Combs and S.A. McCommas. 1986b. Morphometric differentiation among experimental lines of the housefly in relation to a bottleneck. Genetics 114: 1213-1223.

Bryant, E.H. and L.M. Meffert. 1990. Fitness rebound in serially bottlenecked populations of the housefly. American Naturalist 136: 542-549.

Bryant, E.H. and L.M. Meffert. 1992. The effect of serial founder-flush cycles on quantitative genetic variation in the housefly. Heredity 70: 122-129.

Bush, G.L. 1975. Modes of animal speciation. Annual Review of Ecology and Systematics 6: 339-364.

Bush, G.L. and S.R. Diehl. 1982. Host shifts, genetic models of sympatric speciation, and the origin of parasitic insect species. Proceedings of the 5th International Symposium om Insect-Plant Relationships, Wageningen.

Carson, H.L. 1975. The genetics of speciation at the diploid level. American Naturalist 109: 83-92.

Carson, H.L. and A. Templeton. 1984. Genetic revolutions in relation to speciation: the founding of new populations. Annual Review of Ecology and Systematics 15: 97-131.

Carson, H.L. 1990. Increased genetic variation after a population bottleneck. Trends in Ecology and Evolution 5: 228-230.

Coyne, J.A. 1993. The genetics of an isolating mechanism between two sibling species of Drosophila. Evolution 47: 778-788.

Dobzhansky, T. 1955. Evolution, Genetics, and Man. Wiley, New York, USA.

Dobzhansky, T. 1962. Mankind Evolving.Yale University Press, Boston, USA.

Feder, J.L., C.A. Chiloe and G.L. Bush. 1988. Genetic differentiation between sympatric host races of the apple maggot fly Rhagoletis pomonella (Diptera: Tephritidae). Nature 336: 61-64.

Feder, J.L. and G.L. Bush. 1989. Gene frequency clines for host races of Rhagoletis pomonella in the Midwestern United States. Heredity 63: 245-266.

Feder, J.L. 1995. The effect of parasitoids on sympatric host races of Rhagoletis pomonella (Diptera: Tephritidae). Ecology 76: 801-813.

Feder, J.L., U. Stolz, K.M. Lewis, W. Perry, J.B. Roethele and A. Rogers. 1997. The effect of winter length on the genetics of Rhagoletis pomonella (Diptera: Tephritidae). Evolution 51: 1862-1876.

Fry, J.D. 2003. Multilocus models of sympatric speciation: Bush versus Rice versus Felsenstein. Evolution 57: 1735-1746.

Graur, D. and D. Wool. 1982. Dynamics and genetics of mating behaviour in Tribolium castaneum (Coleoptera: Tenebrionidae). Behavior Genetics 12: 161-179.

Haeckel, E. 1876. The History of Creation. Appleton & Co., New York, NY, USA.

Huxley, J. 1942. Evolution, the Modern Synthesis. Belknap Press, Harvard, USA.

Huettel, M.D. and G.L. Bush. 1972. The genetics of host selection and its bearing on sympatric speciation in Procecidochares (Diptera: Tephritidae). Entomologia Experimentalis et Applicata 15: 465-480.

Itami, J.K., T.P. Craig and J.D. Horner. 1997. Factors affecting gene flow between three host races of Eurosta solidaginis. pp. 375-407. In: S, Mopper and S. Strauss (eds.), Local Genetic Structure in Natural Insect Populations: Effects of Host Plant and Life History, Chapman & Hall, New York, USA.

Lyell, C. 1973 (originally published 1863). The Geological Evidence for the Antiquity of Man. AMS Press, New York, USA.

Mayr, E. 1942. Systematics and the Origin of Species. Columbia University Press, New York, USA.

Mayr, E. 1963. Animal Species and Evolution. Harvard University Press, Boston, USA.

McCommas, S.A. and E.H. Bryant. 1990. Loss of electrophoretic variation in serially bottlenecked populations. Heredity 64: 315-321.

Mivart, St.George J. 1871. The Genesis of Species. MacMillan, London, UK.

Powell, J.R. 1978. The founder-flush speciation theory: an experimental approach. Evolution 32: 465-474.

Prokopy, R.J., S.R. Diehl and S.S. Cooley. 1988. Behavioral evidence for host races in Rhagoletis pomonella flies. Oecologia 76: 138-147.

Smith, S.G. 1952. The evolution of heterochromatin in the genus Tribolium (Coleoptera: Tenebrionidae). Chromosoma 4: 585-610.

Templeton, A.R. 1981. Mechanisms of speciation: a population genetic approach. Annual Review of Ecology and Systematics 12: 23-48.

Wagoner, D.E., C.A. Nickel and O.A. Johnson. 1969. Chromosomal translocation heterozygotes in the housefly. Journal of Heredity 60: 301-304.

Waisel, Y. 1959. Ecotype variation in *Nigella arvensis*. Evolution 13: 469-475.

Wool, D. 2001. Charles Lyell, "the father of geology", as a fore-runner of modern ecology. Oikos 94: 385-391.

CHAPTER 23

Speciation, Extinction of Species and Phylogeny

INTERPRETATIONS OF THE FOSSIL EVIDENCE: CUVIER, LYELL AND DARWIN

Palaeontologists studying the fossil record found no evidence for a gradual process of speciation. New fossil species appeared suddenly in different geological strata, and just as suddenly disappeared without trace.

Georges Cuvier and Charles Lyell were aware of the disappearance of fossil species when geological formations were compared. Lyell in particular was in doubt as to whether some of the missing species did still exist somewhere in unexplored parts of the earth. The disappearance of the large reptiles, which had ruled the earth in the Miocene, was certainly a case of extinction: it was unlikely that animals as large as the Megatherium (described by Cuvier) could still be alive somewhere without being noticed.

Cuvier thought that these extinctions were due to catastrophes, which periodically destroyed all life on earth. Lyell believed that extinction resulted from slow, gradual environmental changes that were continuously taking place: the ordinary factors of volcanic activity and erosion were sufficient to bring about changes in the habitat and, consequently, the plant and animal species living at a site would be replaced by others. Lyell thought that transmutation of one species to another, as suggested by Lamarck – even if it were possible – would be an extremely slow process. A much faster process of species replacement was migration from other areas, where species adapted to the new conditions already existed.

Darwin thought of a continuum, a gradual transition in nature, from populations through subspecies and finally good species. He was aware of the difficulty inherent in the theory of descent: if species descended gradually and with modifications from ancestors, many fossils should have been found showing transitional stages between past and present species. This is not the case. Fossils species are often clearly distinct from each other and from extant forms. New species seem to have appeared at once, with no intermediates.

Darwin lists a number of reasons for this discrepancy – in particular, the incomplete geological record. His explanations seem reasonable to people who believe in evolution. But Creationists still use the gaps in the geological record as evidence against the theory.

The idea that these gaps were formerly occupied by currently-missing intermediate forms was rejected by Goldschmidt (1940):

> There is, I think, in the whole idea of subspecies as incipient species a psychological element. It is taken for granted that species evolve from each other by slow accumulation of small individual steps (by means of selection, of course)… If, nevertheless, the individual rassenkreise remain separated by large gaps… the pre-concieved idea forces the neo-Darwinist to look for the most impossible explanations to fill the gaps… One of these which always works is that the existing gaps were formerly filled with missing links. In other words, the subspecies are incipient species because a strictly Darwinian view <u>requires</u> such an interpretation (Goldschmidt 1940, pp. 136-7. Review in Dietrich, 2003).

Darwin was not surprised by extinction. Old species could become extinct not only by disappearing from the face of the earth, but also when they changed form (by natural selection) and became other species. In the former case, however, he thought that extinction was a result of competition between species, the less-well-adapted being removed by better-adapted species. He had in mind a process of group selection, working among species: extinction is the fate of small and diminishing groups of organisms, which must be less fit than larger groups:

> The largeness of any group shows that its species have inherited, from a common ancestor, some advantage in common… Within the same large group, the later and most highly perfected subgroups…will constantly tend to supplant and destroy the earlier and less-improved subgroup. Small and broken groups and subgroups will finally disappear (Origin, p. 275 ff).

MASS EXTINCTIONS: METEORITES AND CUMULATIVE DESTRUCTION BY MAN

As more palaeontological data accumulated, scientists discovered evidence for simultaneous, mass extinction of many groups, which could only be explained by an external causative factor – an explanation closer to Cuvier's catastrophes than to Darwin's view of gradual extinction by competition (See Price, 1996 for a more detailed account). A very large meteorite colliding with the earth is thought to have caused some of the major mass extinctions. Craters have been found on the surface of the globe (e.g. in Arizona, USA, and in Mexico) which could perhaps document such collisions. Giant meteorite collisions could have caused major climatic and other environmental changes that destroyed many taxa simultaneously.

Extinction of species is today a major concern for conservationists and ecologists interested in species diversity. The extinction of species by deliberate human activity is not a new problem: Lyell in the early 19th century was aware that

> Man is, in truth, continually striving to diminish the natural diversity of the stations of animals and plants, in every country, and to reduce them all to a small number fitted for species of economic use (Lyell, 1853, p. 682).

Man destroys the habitats suitable for other organisms – e.g., by deforestation, a process which still goes on in the virgin forests of Indonesia and the Amazon forests in Brazil, despite the protests of scientists – to prepare more land for agriculture. Deforestation in Europe caused the replacement of forests by swamps (Lyell, 1859). The native mammalian fauna of Africa has been destroyed by excessive hunting (see Price, 1996 for a recent review). The most important cause of extinction today is the fast destruction of habitats available for wild organisms (e.g., Lande, 1998). And even conservationists tend to ignore the fact that many species of small animals, particularly insects, are disappearing at an alarming rate, even before their existence has been recorded (Dunn, 2005).

GRADUALISM VERSUS 'PUNCTUATED EQUILIBRIA'

Palaeontologists who follow the tracks of evolution as it appears in the geological fossil record fail to find evidence for a continuum of transitional stages between species. Many new forms often appear in a short (geological) time, to be followed by a long period of 'stasis' when species seem to remain unchanged, until the next 'burst' of speciation events. For example, all recent groups of Mammalia, from shrews to mammoths, first appeared within a short time in the Miocene, and there is no evidence of further speciation after that time.

Another fact, which is not in line with gradual change, is the presence of 'living fossils' – species found living today, but known as fossils from remote periods, with no noticeable difference in morphology between extinct and extant forms. Why these organisms have remained unchanged is not clear:

perhaps their environment remained stable for eons, or their populations lacked the genetic variation necessary for evolution by natural selection. These ideas are not entirely satisfactory. For example, one living fossil - the horseshoe crab (*Limulus*) – still lives in the same environment with other crustaceans, which did evolve, as the fossil record shows. *Limulus* is no less variable electrophoretically than other crabs (Selander et al., 1970).

Gould and Eldredge (1977) believed that new species are formed by 'saltations', and that the fossil record is the true description of the process of macro-evolution (their term for this model: <u>punctuated equilibria</u>) (Fig. 23.1). They claim that gradualism is imposed by Darwin's theory, and is not a fact of nature. As suggested by Goldschmidt, they argue that evolution of a new species is not due to slow accumulation of mutations by natural selection, but requires a different mechanism. Macro-evolution is not an extention of micro-evolution.

Phyletic gradualism and punctuated equilibria are widely different concepts of evolution and imply quite different processes.

    (1) <u>Rates of evolution</u>: Gradualism implies a constant rate of accumulation of mutations. Punctuation implies that rates of evolution are variable: short bursts of rapid change followed by long periods of no change at all (stasis).

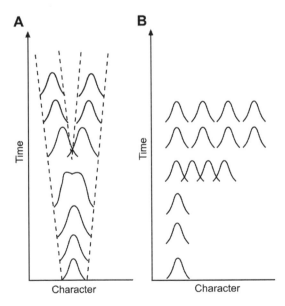

Fig. 23.1 Models of speciation rates in evolution: gradualism (left) and punctuated equilibria (right).

(2) <u>Population size</u>: Gradualism implies that speciation (macro-evolution) should occur in large, variable populations. According to Fisher's Fundamental Theorem (Chapter 8) evolutionary change will be faster, if the population is more variable. In the punctuation model, evolution is more likely to occur in small, isolated populations.

(3) <u>Directionality</u>: Gradualism implies a more-or-less continuous change in the same direction.The punctuated equilibrium model implies that during the speciation episode, several different lines may proceed in different directions.

(4) <u>Genetics and ecology</u>: Gradualism implies a genetic process of slow accumulation of mutations, which is independent of the environment. According to the punctuation model, the burst of speciation events coincides with the presence of environmental niches available for colonization (Gould and Elderdge, 1977).

The controversy over the two models continued for some years. Hoffmann (1982) supported the punctuated equilibrium theory. He suggested that reproductive barriers may form 'randomly' in small populations and create new species, some of which may survive. Hoffmann mentioned also that several difficult biological problems can be solved if the possibility of 'species selection', as a separate process from individual selection within populations (Stanley, 1975), is allowed. For example, following this approach, evolution in asexual populations – where the 'biological species' definition does not apply – is no longer problematic.

These arguments failed to convince other evolutionists. Charlesworth et al. (1982) argue that the terminology of the 'punctuated equilibrium' is unnecessary, and does not contradict the Darwinian scenario for speciation. They argue that there is no evidence that evolutionary change occurs especially in small and isolated populations. It seems (to these authors) impossible to relate speciation events – i.e., the formation of reproductive barriers – to quantitative morphological change in fossils: since it is impossible to test for reproductive barriers in fossils, the decisions about speciation and species boundaries are based entirely on morphology. In living sibling species, absolute reproductive barriers exist without any morphological change. Charlesworth et al. also reject the idea of 'species selection' as a different mechanism from individual selection.

EVOLUTION AS THE HISTORY OF LIFE (PHYLOGENY)

The idea that all forms of life are connected by descent from primitive, spontaneously-generated ancestors (monads) was introduced in biology by Jean Baptiste Lamarck in 1809. This 'theory of descent' (Haeckel, 1876) is the basis of the description of the evolution of life as a historical branching tree (**phylogenetic tree**), first attempted by the German biologist, Ernst Haeckel.

Haeckel described all forms of animal life as descended from a primitive ancestral form which he called **gastrea**. This organism had just two layers of cells surrounding a central digestive cavity, like the coelenterates of today. One of the early embryological stages of the vertebrates, called <u>gastrula</u>, has a similar structure. This was, for Haeckel, evidence for the community of descent of invertebrates and vertebrates. Haeckel went further and constructed detailed branching trees for the vegetable kingdom, and for the major groups within the animal and plant kingdoms – supplanting the scant evidence then available with his great artistic talent and imagination (Fig. 23.2).

Phylogeny is a veteran and rich field of study in evolution. It combines data on extant organisms with the fragmentary fossil evidence and reconstructs the possible pathways of evolution from ancestors to the present-day species. The difficulty is that, although extant organisms and fossils provide solid evidence, the lines that connect them are only 'educated guesses' representing the conclusions of trained scientists based on the currently available evidence (Fig. 23.3). The drawings of phylogenetic trees in textbooks and scientific articles are of course tentative and changeable: they are snapshots of available knowledge.

Some of the assumptions on which phylogenetic trees are based are the following (Kerkut, 1960):

(1) Life first originated from non-living material.

(2) There was only one origin to all life (monophyletic origin)

(3) All organisms – viruses, bacteria, plants and animals – are connected by descent.

(4) Multicellular organisms evolved from unicellular predecessors.

(5) All groups of invertebrates are connected by descent.

(6) Vertebrates descended from invertebrate ancestors.

(7) Within the vertebrates, amphibia descended from fishes, reptiles from amphibia, and the birds and mammals descended from reptiles.

The assumption of monophyletic origin (2) is reflected in all of Haeckel's phylogenetic trees. An alternative, polyphyletic (bush-like) structure is preferred by some scientists. The strongest evidence in support of assumption (2) is, of course, the established similarity of the

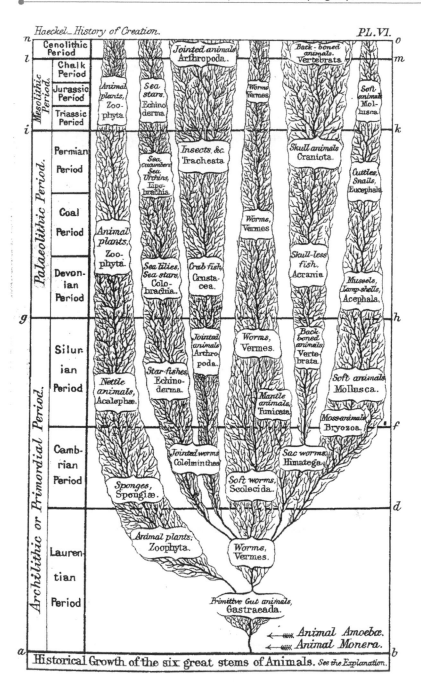

Fig. 23.2 Phylogeny of the major animal taxa (from Haeckel, 1876).

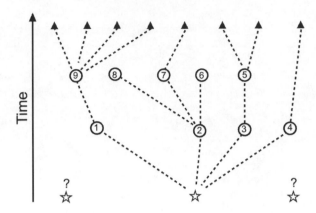

Fig. 23.3 Schematic phylogeny. Triangles represent real species. Numbered circles are known fossils (#4 may be a living fossil, #6 is a fossil representative of a lineage now extinct). The lines represent one of many possible evolutionary pathways. Stars point to the unknown origin(s?) of life.

genetic code throughout the animal kingdom, which enabled the comparisons of DNA sequences from different organisms and the calculation of evolutionary rates in different lines. Kerkut (1960) thought that supporters of monophyly tend to ignore evidence to the contrary: the classification of organisms, founded by Linnaeus in the 18th century, is based on morphological similarity, and ignores differences in other traits – like metabolic pathways which diverge within groups. The enzymes involved in DNA polymerization, duplication and repair do not have to be the same although the DNA code is similar (Kerkut, 1960).

THE DATA

Phylogenetic studies necessarily derive most of the data from extant organisms. Darwin interpreted the similarity between different species as indicating community of descent: the more closely related two species are, the greater the similarity between them. In building phylogenetic trees, scientists join together organisms that seem to be similar to each other, accepting Darwin's interpretation of similarity as a measure of community of descent.

In Darwin's and Haeckel's time the evidence came from comparisons of anatomical, morphological and embryological characters. Recently, estimates of similarity – and, by inference, community of descent – come mostly from comparison of DNA sequences in different organisms, enabled by the advances in molecular techniques (review in Graur and Li, 2000).

The phylogenetic trees resulting from molecular data do not always agree with accepted trees based on morphology, raising the (unanswerable) question of which is closer to the true, unknown phylogeny.

The earliest fossil of something that looks like a cell is from the pre-Cambrian rocks of South Africa, dated about 3 billion years ago. Since this specimen seems to have a clear cell wall, it must have been preceded by millions of more primitive ancestors of which no traces have been discovered. It is commonly accepted that the first organisms were heterotrophic, feeding on the organic materials surrounding them in the 'primeval soup'. The autotrophic organisms (green organisms, containing chlorophyll) are thought to have evolved later: although they require only minerals, CO_2 and sunshine from the environment, the complex biochemical machinery needed for photosynthesis cannot be expected in the primitive stages of life. Once this machinery evolved, however, the free oxygen released in photosynthesis accumulated in the atmosphere, and enabled aerobic metabolism, which is much more efficient in storing energy (in the form of energy-rich ATP) than anaerobic metabolism. The evolution of larger, multicellular organisms was thus made possible.

Among the Flagellata – unicellular green organisms with a flagellum – there are organisms which look like groups of 4, 8, 16, 32, or 64 cells each with its own flagellum, joined together. Then there is Volvox, a spherical organism with thousands of cells. This sequence can be used as a model for the evolution of multicellular from unicellular organisms, if the dividing cells remain attached to each other rather than separate. (It should be kept in mind, however, that flagellates are autotrophic and are not considered the primitive forms of organisms.)

Working out the phylogeny is easier in vertebrates than in invertebrates. The great anatomical similarity between fishes, amphibians, reptiles, birds and mammals, was noticed by anatomists in the 19th century, before Darwin. These similarities are further supported by evidence from the embryological development of these organisms, of which Darwin was well aware. But for Georges Cuvier and Etienne Geoffroy saint-Hilaire in the Natural History Museum in Paris, who developed the science of comparative anatomy, the similarity meant only that these animals belonged in the same taxonomic group: they totally rejected the idea of common descent suggested by their colleague, J.B. de Lamarck.

Fossils are, of course, very helpful in the construction of phylogenies. The amount of available palaeontological evidence varies greatly among groups of organisms. Soft bodied animals that lack a bony skeleton or a hard shell are unlikely to leave a trace in the rocks. On the other hand, mollusks and echinoderms are very common as fossils. The formation of fossils also depends on the conditions at the site where the animal and plant remains sank into the mud in some ancient lake. Darwin's statements about the reasons for the 'incompleteness of the geological record' are certainly justified.

Great help in the reconstruction of phylogeny is provided by rare fossil finds like the skeletons of *Archaeopteryx*, a fossil reptile with bird-like characters which lived about 180 million years ago and was discovered in the 1860's (Wellnhofer, 1990). It had wings with feathers and probably could glide between trees or even fly. This fossil provided a possible link between reptiles and birds.

PHENETICS, CLADISTICS, AND PHYLOGENY

Numerical Taxonmy (Chapter 21) uses the observed (phenotypic) characters of the organisms or taxa (Operational Taxonomic Units (OTUs) in NT terminology) as data. NT gives equal weight to all characters, and makes no assumption about descent (Sokal and Sneath, 1963; Sneath and Sokal, 1973). This approach is called **phenetics**. It is argued that many characters, of all kinds, should be measured on each OTU – whether morphological, electrophoretic, chemical or behavioral characters – ideally on many individuals for each taxon and on different life history stages – for the similarity between OTUs to be close to the true relationship between them. The resulting dendrogram – called a phenogram – is not intended to reflect similarity by descent, although if one accepts Darwin's interpretation that morphological similarity between taxa is a result of common ancestry, then the phenogram can be a reflection of the pattern of descent.

A rival theory, called **cladistics**, branched off from Numerical Taxonomy. The cladistic philosophy is that similarities among taxa should reflect the phylogenetic relationships among them. Accordingly the cladistic methodology claims to classify organisms by their community of descent.

The database is assembled as in the phenetic NT studies, but not all of the collected measurements can be used as characters in cladistics. A number of unprovable assumptions, actually taken for granted and treated as axioms, are introduced at the stage of data analysis. It is up to the investigator to select the 'informative' characters.

One of the basic assumptions in character selection is that each group of two or more closely similar taxa had a monophyletic origin (forming a **clade**). A second assumption is that it is possible to determine which state of each character is primitive (ancestral), and which are derived states. For example, the absence of functional eyes in cave-dwelling organisms is a derived state, if one assumes that they evolved from free-living organisms with functional eyes.

No ancestral state of any character is actually known, but it can be inferred from the phylogenetic relationship of the taxa. The technique for deciding on ancestry of character states – that will give directionality to the resulting tree, here called a **cladogram** – is the use of an **outgroup**. For example, to study the cladistic relationships of gall-inducing aphids (Fordinae, Pemphigidae) we added a species of non-galling, free-living aphid to the studied taxa (it can be safely assumed that galling aphids are derived from non-galling ones, and not the other way around (Inbar et al., 2004). The character states of the outgroup species are treated as ancestral (more primitive) states (Maddison, 1994).

The numbers of trees that can be constructed by the sophisticated cladistic algorithms from data of extant organisms quickly becomes enormous as the number of species (OTUs) increases. For example, with three taxa, there are three possible ways to connect them. To connect four taxa, there are 15 possible trees. The number of possible trees for 10 taxa – still a small number in phylogenetic studies – is more than 34 million! The number of rooted trees can be calculated from the relation

$$N_r = (2n - 3)! / 2^{n-2} (n - 2)!$$

where n is the number of taxa in the study, and ! (factorial) stands for the product of all integers from n to 1 [5! = 5*4*3*2*1] (Li and Graur, 1991; Graur and Li, 2000).

It is important to emphasize that each group of taxa **can have only one true phylogeny,** which is unknowable. There is no way of knowing which of the resulting trees is the closest to the true phylogeny. Computer algorithms that choose the 'best' trees use the criterion of **parsimony**: a tree requiring the smallest number of steps (estimated mutational changes) from the ancestor (outgroup) to all the tips of the branches (taxa) is considered the most likely; but there is no compelling reason to think that nature, or evolution, is at all parsimonious. In many cases, several equally parsimonious trees are found, with different topologies, and it is up to the investigator to choose between them. Figure 23.4 illustrates the 'true' phylogenetic relationships of seven imaginary species of beetles, which differ in one character: color. The character has three states: white, grey and black. To reconstruct the cladistic relationships, decisions about the ancestral states are required at six branching points. The number of possible trees in this simple cladogram is $3^6 = 729$ (Maddison, 1994).

(In this example, four equally-parsimonious trees were found – each with three steps – although the arrangement of the species at the top of the tree was of course not the same.)

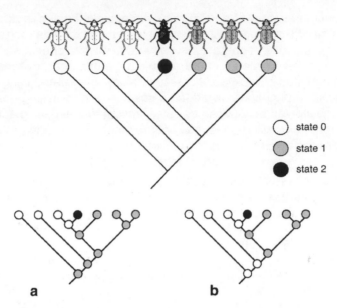

Fig. 23.4 A hypothetical example of cladistic relationships (from Maddison, 1994; Reproduced with permission of the Annual Review of Entomology). The smaller diagrams represent two of the 729 possible trees. The most remote ancestor is assumed to be grey in (a), and white in (b). The number of mutational steps needed to get to the 'true' tree is four in (a), and three in (b). Using the rule of parsimony, (b) is the better tree in this example.

DEDUCING PHYLOGENY FROM MOLECULAR SEQUENCE COMPARISONS

The cladistic methodology was enthusiastically adopted by taxonomists, who felt that classification should reflect evolutionary (phylogenetic) relationships among taxa. Even more enthusiastically, the method was adopted by investigators working with molecular data. Sequence comparisons of a given DNA segment from different organisms provide large numbers of characters for comparison (base differences), and outgroups can be selected by consulting the phylogeny of the organisms in question. The many DNA sequences in data banks are very useful for molecular evolutionists. Building of molecular phylogenetic trees has been very popular in the last two decades (Graur and Li, 2000).

Still, the sequences available for comparison - of a very large number of genes – come from a limited number of organisms, representing the few taxa which were of interest for humans; Apart from man himself, sequences are available from some primates, a few other mammals (horse, cow, pig, rabbit, rat and mouse), *Drosophila*, and some bacteria and fungi. Anyone interested in the phylogeny of other groups, needs to do the sequencing himself – a laborious and still expensive procedure. Within taxa, the number of individuals used is very small – no way near the numbers that were studied for electrophoretic variation at the population level.

Molecular phylogenies must rely on conventional, organismal phylogeny and palaeontology for estimating the time scale and the divergence points of large groups. Molecular phylogenies do not always have the same topology as the conventional ones. For example, the marine mammals (dolphins and whales) have distinct adaptive characters for living in the sea, which set them clearly apart – morphologically and therefore taxonomically – from land animals. But protein and mtDNA sequences place them close to the hooved herbivores (Artiodactyla) (Graur and Higgins, 1994).

An interesting case of such a conflict, which was published not only in scientific journals but also in the daily press, was the allocation of the cavia, a South American mammal and common pet ('guinea pig' is a misnomer, since it has nothing to do with pigs nor does it live in Guinea) – to a primitive mammalian group at the base of the mammalian phylogeny, despite its morphological similarity to rodents (Graur et al., 1991; Li et al., 1992).

PROGRESS AND TRENDS IN PHYLOGENY

Palaeontologists sorting a collection of fossils from consecutive strata over geological time have long noticed trends. Some of these trends were observed by Lamarck, like the progressive development "from the simple to the more complex". One of the clearest trends is the observed increase in animal size, as illustrated frequently in textbooks by the evolution of the modern horse – from dog-sized Eohippus to the modern animal. (see Fig. 13.3). Are these trends real – or artifacts of sampling and imagination? And if real, what brings them about?

There is no doubt among evolutionists that the first organisms were microscopically small ('monads', in Lamarck's terminology). There is also no doubt that later, multicellular organisms were both larger and more complex. Can the increase in size be attributed to some selective advantage of larger forms?

In some cases, large size may be advantageous: a larger predator may catch more prey, or alternatively, larger prey may be more successful in escaping predators. Large size in homoeothermic organisms may be more efficient in energy conservation in cold climates (there seems to be a lower limit to small size in mammals – the smallest shrew weighs about 3 g – while in poekilotherms like insects, much smaller animals exist).

While an increase in complexity may be achieved without disrupting the normal functions of the organism (by adding some new organs), a reduction in complexity means loss of organs and can cause such disruption (Arthur, 1984). This can give a selective advantage to an increase of complexity.

Stanley (1973) showed with palaeontological data that other explanations are possible. The earliest fossil individuals in all groups were small, and later ones larger. Using computer simulation, he showed that if there is a minimal size, below which existence is impossible, a trend of size increase will emerge because the minimum is a limiting boundary while the maximum is not (Gould, 1988). Later species will be equal to or larger than their predecessors, moving the mean to the right. There is no need to postulate selective advantage.

THE CAMINALCULUS

To select the most likely of alternative phylogenetic trees, or to compare methods of constructing them, one needs to know the true tree. In nature, the true phylogeny is unknowable. The problem was simulated in a study of the phylogeny of imaginary organisms, created by the late Prof. Joseph Camin, of the Department of Entomology at the University of Kansas, and given their 'scientific' name Caminalculus, after him (Sokal and Rohlf, 1980; Sokal, 1983a, b).

Starting with a 'primitive' form, Camin created an entire phylogenetic tree for 77 'taxa', by tracing a drawing of a former 'taxon' with modifications to create a new one ('descent with modification'; Fig. 23.5).

Camin followed rules and principles that are used in animal phylogeny: some lines of descent were 'dead ends', ending in extinction. Others were 'living fossils' that did not change at all along the 'time' scale, and still others speciated at different time points. The 'recent fauna' of Caminalculus comprised 29 forms, differing in shape and in the pattern of black patches on their 'bodies' (originally there were 30, but two of them were created exactly alike in order to see whether their identity would be recognized. It was).

The true phylogenetic tree was drawn by the artist-creator and kept secret in his safe. The drawings of the 29 'recent' Caminalcules (Fig. 23.6) were given to different individuals, who were asked to group the forms and reconstruct their phylogeny – and to state the rules and reasons for their decision about the descent of each form. Among the individuals who examined the drawings were university professors, graduate students of entomology, technicians, and even children. A random examination of the drawings was simulated by placing 25 computer cards, with 25 random

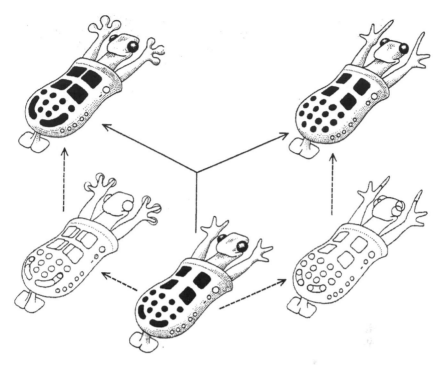

Fig. 23.5 An illustration of descent with modifications in the phylogeny of Caminalculus (from Sokal, 1966; Reproduced with permission of the publisher)

holes punched in each (readers may not remember this early stage of data storage in computer technology, which was still in use only 35 years ago!) on each of the 29 drawings. For each of the holes, a score of 1 was recorded when a black line was visible, otherwise it was scored as 0 – for a total of 625 data points. The matrix of 29 × 625 1s and 0s was then analyzed by computer and the similarity between the 29 OTUs was estimated.

The use of these imaginary organisms, instead of real animals, in studies of phylogenetic methods has two advantages. First, all the available information is contained in the two-dimensional paper drawings – a purely morphological set of data. There are no books to consult, no data on the animal's physiology, genetics, reproductive isolation or behavior which could affect the decisions about their descent. The observer, however, is not free of concepts of animal evolution – derived from former studies or from familiarity with real organisms – and these concepts affect his/her decisions. A taxonomist who examines dead, preserved specimens and decides their species status and affinities (assuming the Biological Species

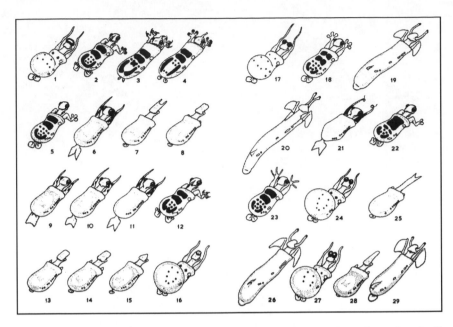

Fig. 23.6 The 29 surviving OTUs of Caminalculus (from Sokal, 1983a; Reproduced with permission of Taylor and Francis Group).

definition without ever seeing the organisms alive) is faced with a very similar situation. His decisions are based, perhaps unconsciously, on his knowledge of biology and his pre-conceptions about speciation.

The second, important advantage of the Caminalculus is, of course, that the true phylogeny is now known (published by Sokal, 1983a, b) and the results of different methods of reconstructing phylogeny can be compared to the true tree.

Author's Comment

I used these drawings in laboratory sessions in my Ecological Genetics courses at Tel Aviv University. Graduate students were given cards with the 29 'recent' images of Caminalculus and asked to reconstruct their phylogeny. They worked in groups of three, and free discussions and arguments were encouraged. I highly recommend such sessions, which illustrate to the students – much better than any number of frontal lectures – to what degree their decision on classification and phylogeny are based on their biases and preconceptions rather than on the data. From the start, terms like 'head', 'legs', and 'eyes' were invariably used for communicating with each other, although all that the students had to deal with were two-dimensional line drawings. One example, which invariably was a source of argument, was the question: does evolution of a given character – say, the number of eyes – proceed from the simple to the more complex - or the other way around, like the case in cave-dwellers? This is difficult enough to decide in real organisms, let alone in the Caminalculus where you do not know where the 'organisms' live – actually, there is no guarantee that the structures you observe are eyes at all!

In the advanced stages of the laboratory sessions I offered the students 'new palaeontological finds' (drawings of some of the 48 extinct forms in the Caminalculus phylogeny), one at a time, to see how these improved their reconstruction of the phylogeny. No one succeeded in getting the entire tree right – even given all the 'fossils' (but not telling them where the 'fossil' forms belonged) – a reminder that the exact reconstruction of the sequence of past events may be impossible.

Literature Cited

Arthur, W. 1984. Mechanisms of Morphological Evolution. Wiley & Sons, New York, USA.

Charlesworth, B., R. Lande and M. Slatkin. 1982. A neo-Darwinian commentary on macroevolution. Evolution 36: 474-498.

Darwin, C. 1950 (1859, 1871). The Origin of Species/the Descent of Man. Modern Library, New York, USA.

Darwin, F. 1887. Life and Letters of Charles Darwin. Murray, London, UK.

Dietrich, M.R. 2003. Richard Goldschmidt: Hopeful Monsters, and other "heresies". Nature Reviews – Genetics 4: 68-74.

Dunn, R.R. 2005. Modern insect extinctions: the neglected majority. Conservation Biology 19: 1030-1036.

Goldschmidt, R. 1940. The Material Basis of Evolution. Yale University Press, Boston, USA.

Gould, S.J. and N. Eldredge. 1977. Punctuated equilibria: the tempo and mode of evolution reconsidered. Paleobiology 3: 115-151.

Gould, S.J. 1988. Trends as changes in variance: a new slant on progress and directionality in evolution. Journal of Palaeontology 62: 319-329.

Graur, D. and W-H. Li. 2000. Fundamentals of Molecular Evolution. Sinauer, Sunderland, USA.

Graur, D., W.A. Hide and W-H. Li. 1991. Is the guinea-pig a rodent? Nature 351: 649-651.

Graur, D. and D.G. Higgins. 1994. Molecular evidence for the inclusion of Cetaceans within the order Artiodactyla. Molecular Biology and Evolution 11: 357-364.

Haeckel, E. 1876. The History of Creation. Appleton, New York, USA.

Hoffman, A. 1982. Punctuated versus gradual mode of evolution: reconsideration. Evolutionary Biology 15: 411-430.

Inbar, M., M. Wink and D. Wool. 2004. The evolution of host-plant manipulation by insects: molecular and ecological evidence from gall-forming aphids on *Pistacia*. Molecular Phylogenetics and Evolution 32: 504-511.

Kerkut, G.A. 1960. Implications of Evolution. Pergamon Press, London, UK.

Lande, R. 1998. Anthropogenic, ecological and genetic factors in extinction and conservation. Researches in Population Ecology 40: 259-269.

Li, W-H., A. Zharkikh, D.P. Ma and D. Graur. 1992. The molecular taxonomy and evolution of the guinea pig. Journal of Heredity 83: 174-181.

Lyell, C. 1853. Principles of Geology. (9th ed.) Murray, London, UK.

Maddison, D.R. 1994. Phylogenetic methods for inferring the evolutionary history and processes of change in discretely valued characters. Annual Review of Entomology 39: 267-292.

Price, P.W. 1996. Biological Evolution. Saunders College Publications, New York, USA.

Selander, R.K., S.Y. Yang, R.C. Lewontin and W.E. Johnson. 1970. Genetic variation in the horseshoe crab (*Limulus polyphenus*), a phylogenetic 'relic'. Evolution 24: 402-414.

Sneath, P.H.A. and R.R. Sokal. 1973. Numerical Taxonomy. Freeman, New York, USA.

Sokal, R.R. 1966. Numerical taxonomy. Scientific American 215: 106-116.

Sokal, R.R. 1983a. A phylogenetic analysis of the Caminalculus. I: the data base. Systematic Zoology 32: 159-201.

Sokal, R.R. 1983b. A phylogenetic analysis of the Caminalculus. II: Estimating the true cladogram. Systematic Zoology 32: 185-201.

Sokal, R.R. and P.H.A. Sneath. 1963. Numerical Taxonomy. Freeman, New York, USA.

Sokal, R.R. and F.J. Rohlf. 1980. An experiment in taxonomic judgement. Systematic Botany 5: 341-365.

Stanley, S.M. 1973. An explanation of Cope's Rule. Evolution, 27: 1-26.

Stanley, S.M. 1975. A theory of evolution above the species level. Proceedings of the National Academy of Sciences, USA, 72: 646-650.

Wellnhofer, P. 1990. Archaeopteryx. Scientific American 262: 70-77.

CHAPTER 24

Evolutionary Processes in Human Populations

Though biological science was content to classify him [man] as just another animal, in his own eyes he still is the Lord of Creation, apart from the rest of nature, and in some unspecified sense, above nature (Huxley, 1957, p. 42)

Man is considered the crown of Creation. Of course, he awarded that title to himself. He also considers himself the ruler of the entire living world. And even this is in serious doubt. Because, even if he rules the horse, the cow, and the dog, this does not mean that he rules, for example, the mosquitoes and the fleas, which bite him to death and do not even let him sleep peacefully" (Avigdor Hameiri, "Animal Wisdom" (1933), translated from Hebrew).

HISTORICAL NOTES

Descended from the apes! My dear, we hope that it is not true. But if it is, let us hope that it may not be generally known (Dobzhansky 1961, p. 5)

No other chapter of evolution has been as controversial as the evolution of man.

In the 18th century, Linneus included man in the same group, Anthropomorpha, with the primates in the system of nature (Systema Naturae) which he thought represented the Divine plan of the world. Linneus did not claim that man was <u>descended</u> from the same stock as the primates, only that he was similar to them – and that this was how the Creator had designed him. No objections seem to have been raised to this classification.

In the Origin of Species, in 1859, Darwin did not disclose his thoughts about the origin of man, other than the vague statement that "light will be shed" on his origin by the theory of natural selection. But his views were no secret much before he published them, clearly and unambiguously, twelve years later in the Descent of Man (1871):

Thus we can understand how it has come to pass that man and all other vertebrate animals have been constructed on the same general model, why they pass through the same early stages of development, and why they retain certain rudiments in common. Consequently we ought frankly to admit their community of descent... It is only our natural prejudice, and that arrogance which made our forefathers declare that they were descended from demi-gods, which leads us to demur to this conclusion (Darwin, 1874).

The idea that man has descended from the apes upset many at the time, including some leading contemporary biologists such as the anatomist Richard Owen, although Darwin wrote clearly that no one of the existing primates is a direct human ancestor, only that man and ape had a common ancestor in the remote past. The Bishop of Oxford, Samuel Wilberforce, had this to say:

Man is not, and cannot be, an improved ape. Man is the sole species of his genus, the sole representative of his Order and Subclass. Thus I trust (says Owen) has been furnished the refutation of the notion of a transformation of an ape into the man.

Now we must say at once, that such a notion is absolutely incompatible not only with single expressions in the word of God on that subject of natural science, but with the whole representation of the moral and spiritual condition of man (Wilberforce, 1860).

Darwin described in detail the anatomical and embryological similarities of man and the vertebrates in the 'Descent of Man' in an attempt to show that man was a product of the same natural selection that had shaped the rest of the biological world. He was greatly supported by his friend and ally, Thomas Henry Huxley, who wrote a book on the subject (Huxley, 1900). In a comprehensive review of the biology and the habits of the chimpanzee, the gorilla, the orang-utan and the gibbon, Huxley agreed with Linneus that man belongs in the same order with the primates – but not in the same family:

Man and the man-like apes certainly justify our regarding him (man) constituting a family apart from them, but... there is no justification for placing him in a distinct order...

And thus the sagacious foresight of the great lawgiver of systematic zoology [Linnaeus], becomes justified, and a century of anatomical research brings us back to the conclusion that man is a member of the same order...as the apes and lemurs (Huxley, 1900).

ANATOMICAL EVOLUTION

Two branches of scientific research deal with human evolution. Anthropologists are interested in the evolutionary and comparative development of human culture. Biologists deal with problems of population growth and expansion, reproductive success, survival and competition for food. In human populations, these two fields of study cannot be really separated.

Humans are distinct from contemporary apes in two main anatomical characteristics, which are useful in tracing the early stages in the evolution of the homonids in fossils: in the size of their brains and in walking erect. Brain size can be estimated in fossils by the volume of the skull. Erect walking is reflected in the shape and structure of the pelvis. Although brain size seems to be critically important in the evolution of the hominidae and their departure from the apes (see below), Le Gros Clark (1964) thought that erect posture was the crucial step, and preceded the development of the brain:

> There is reason to suppose that the primary factor which determined the evolutionary segregation of the Hominidae from the Pongidae... was the divergent modification of the limbs in adaptation to erect bipedalism (Le Gros Clark, 1964, p.157)
>
> While a large brain may be accepted as one of the diagnostic characters of the species *Homo sapiens*, it is not a valid criterion of the family Hominidae (ibid., p.11).

Today no serious objection is raised against the descent of man from the apes. But the question of which of the extant primates is the closest to the human line is controversial. Interestingly, the German biologist, Ernst Haeckel, suggested that a common ancestor existed in Africa - before any fossil evidence of human predecessors was discovered. Haeckel predicted that this ancestor was intermediate in structure between man and ape, and named the hypothetical creature 'pithecanthropus' (monkey-man). He was critical of his contemporaries who hesitated because they could not find a 'missing link' to connect man and the apes:

> ...they look too much for "signs"and for special empirical advances in the science of nature. They await the sudden discovery of a human race with tails, or of a talking species of ape "which fills the narrow gap between the two".

At the time of writing, the chimpanzee is considered the closest living relative of man. It seems to be the most intelligent of the apes (individual chimpanzees are able to understand some human language and even use sign language if taught to do so). Comparative biochemical and molecular studies discovered that man and chimpanzee have identical sequences of many proteins and extremely similar DNA sequences of many genes. Most molecular comparisons among primate genes suggest that man had a common ancestor with the chimpanzee more recently than with other apes (Graur and Li, 2000).

NATURAL SELECTION IN HUMAN EVOLUTION

A man consists of some seven octillion ($7* 10^{27}$) atoms, grouped in about ten trillion (10^{13}) cells. This agglomeration of cells and atoms has some astounding properties: It is alive, feels joy and suffering, discriminates between beauty and ugliness, and distinguishes good from evil... How can agglomeration of atoms accomplish any of these things? (Dobzhansky, 1970).

Some investigators believe that natural selection still plays a role in human populations – for example, in the form of selective mortality of carriers of deleterious mutations in early stages of embryonic development, and selective reproductive ability of different human genotypes – despite the lowered infant mortality in 'civilized' countries due to modern medical care. Neel and Schull (1972) argue that natural selection may become more important when the average number of children per family decreases, as in modern families in Europe. However, these authors found very low correlations in numbers of children per family, and other characters associated with fertility, between a mother and her daughters (r = 0.01 to 0.2). The reported values explain at most 4% of the variation. The authors conclude that natural selection has a minor effect on the genetics of human populations.

CULTURAL EVOLUTION

Early in his evolution, natural selection must have been the dominant force shaping the anatomy, morphology, and behavior of man. But at some point new selective forces began to take over – forces that are far less effective in other organisms. The point of departure was, probably, associated with intelligence, technical ability (the ability to make tools and use them), learning ability, and particularly communication within the group (language) – characters of which early traces are noticeable in some primates. From that stage on, the ability to communicate and pass on knowledge to one's fellows **(cultural evolution)** has been the dominant force in the evolution of man. Cultural characters are not genetically heritable – they are totally acquired, free from the control by physical forces of natural selection.

Cultural evolution has a great advantage: while genetically-coded information is transferable only from parents to their direct offspring, and is therefore rather slow to spread in a population – cultural information spreads very fast in many directions: within one population, from one generation to future generations, and among populations and nations – through spoken and written language, and recently by television and the Internet.

Cultural transmission of information is not unknown among animals, although not nearly developed to the level in humans. An often-cited example of a non-genetic transmission of behavior innovation was observed in England. Great tits (*Parus major*) in cities learned to peck open the cardboard or foil tops of the milk bottles left by milkmen on doorsteps, and drink the top cream. The discovery that the bottles contained a food source was probably made by one or a few birds, and spread by imitation in the population. So far as is known, this behavior was not fixed by mutation before the milk-distribution system was changed: bottles are no longer left on doorsteps.

A similar case of an adaptive behavior – apparently with no genetic basis – was observed in the same species of birds in Israel: the birds discovered that the green, inconspicuous galls of the aphid, *Paracletus cimiciformis* (Fordinae, Pemphigidae) – contained something edible (aphids). Pecking at the gall while on the leaf would not have been helpful, as the leaf offers no resistance (Fig. 24.1). The birds learned to remove the leaflet with the gall, carry it to a suitable branch and peck at it to get the reward (Wool and Burstein, 1992). The inventor(s) of this solution must have been few in number. It is unclear how widespread is this behavior in the bird population.

Waddington's (1957) 'epigenetic landscape' model (chapter 10) may be used to explain how small acquired phenotypic changes may become absorbed in the final species form. Waddington suggested that a favorable phenotypic change may be acquired or learned by the individual (and transmitted to offspring by 'cultural' means) – until some genetic change causing the same phenotype appears by chance (the Baldwin effect). If and when this occurs, the new genotype will have an immediate selective advantage.

A.R. Wallace believed that man has freed himself from natural selection, and that only cultural processes affected his evolution (Wallace, 1889, chapter 15). The point of departure was when man's mental facilities began to develop: the mathematical, musical and artistic faculties, and the moral sense, could not have been the products of natural selection or descended from the apes:

> It follows therefore, that from the time when the ancestral man first walked erect, with hands freed from any active part in locomotion, and when his

Fig. 24.1 Great tits collecting aphid galls; a culturally-transferred behavior (from Wool and Burstein, 1992).

brain power became sufficient to cause him to use his hands in making weapons and tools, housing and clothing, to use fire for cooking, and to plant seeds or roots to supply himself with stores of food, the power of natural selection would cease to act in producing modifications of his body – but would continuously advance his mind through the development of its organ, the brain (Wallace 1889, p. 455).

There is agreement that the technical and intellectual ability of man is somehow related to the development of the brain. The development of the brain may have given the early mammals an advantage over the great reptiles, which had relatively very small brains. (But the early mammals were very small animals, and their advantage over the reptiles may have been due to characters other than relative brain size – such as homeothermy.)

All anatomical evidence shows that the apes, and the suspected fossil intermediates in the hominid line, had far smaller brains than contemporary humans. In particular the remarkable increase of the cerebrum characterizes the genus *Homo*.

MENTAL FACULTIES

Dobzhansky (1955) considered that in the evolutionary past of mankind, intellectual ability may have had a selective value: individuals with a better learning ability had an advantage in being able to find and use environmental resources faster than other individuals, to detect and avoid predators faster, and to protect themselves from other environmental hazards. But, apart from these advantages, his unique information-sharing ability enables man to control a wide range of environments and change them to fit his needs. Cultural evolution is the key to all human activities:

> Cultural influences have set up the assumptions about the mind, the body and the universe with which we begin; pose the questions we ask; influence the facts we seek; determine the interpretation we give these facts; and direct our reaction to these interpretations and conclusions (Gould, 1988, p. 319).

HERITABILITY OF TALENTS

Darwin's nephew, Francis Galton, assembled the following data – among a wealth of other, similar information – in his book 'Hereditary Genius' (first published 1864) as proof that genius is heritable:

> Among the 54 known male ancestors, relatives and descendants of J.S. Bach, 46 were professional musicians, and of these 17 were composers of varying degrees of distinction…Granted that growing up in a family of musicians is propitious for becoming a musician, it is still quite unlikely that the genetic equipment of the Bachs had nothing to do with it…

August Weismann (1891), who rejected the idea of inheritance of acquired characters, was skeptical of Galton's conclusion:

> The Bach family shows that musical talent, and the Bernoulli family that mathematical power can be transmitted from generation to generation, but this teaches us nothing as to the origin of such talents (Weismann, 1891, p. 96).

While citing Galton's data, Dobzhansky (1961) notes that the same kind of data may lead to the opposite conclusion, but he does not preclude the possibility that some of the variation in abilities may be heritable after all:

> [However] no musical talent is known among the 156 ancestors and relatives of Schumann. Although the composer was married to a virtuoso pianist, none of their eight children possessed great musical ability. Such exception does not disprove that musicianship is genetically conditioned: they only show that its genetic basis is complex (Dobzhansky, 1961, p. 79).

Intelligence

Eckland (1972) suggested that the genetic potential for human improvement is exhausted, and to cope with the fast rate of technological advance, new human genotypes are needed and would be selected for. He suggested that positive assortative mating plays a role in human populations, because intelligence (as well as academic success) is important in the choice of partners for marriage. These choices increase the relatedness of parents and offspring in cultural talents and increase the homozygosity at loci influencing IQ (Eckland, 1972).

The debate about 'nature versus nurture' in the context of education is beyond the scope of this book. The reader is referred to S.J. Gould's book, The Mismeasure of Man. In the 19th century, attempts were made to equate brain size (actually, skull volume) with intelligence. A physician in Philadelphia, Samuel Morton, collected several hundred human skulls and estimated their brain volume by filling the brain case with mustard seeds (later, lead shot) which were then poured into a graduated cylinder. In this way he 'proved scientifically' that white Americans and Europeans (Caucasians) were more 'intelligent' than Africans. This approach, with all its statistical and other faults, fostered prejudices and racial hatred, but yielded no explanation for 'intelligence' – whatever is meant by this term (see S.J. Gould, 1981).

GENETIC VARIATION AND THE QUESTION OF HUMAN RACES

Let us suppose that a giant collector from Mars visited the earth, made a collection of human beings, and returned to work them up in his Martian museum. He would... certainly come to the conclusion, applying usual taxonomic standards, that he had found a new family, Hominidae, and within this a number of genera, like the white, the black, the very distinct brown, and the yellow man (Goldschmidt, 1940, p. 122).

Race differences are facts of nature, which can, given sufficient study, be ascertained objectively. Mendelian populations of any kind, from small tribes to inhabitants of countries and continents, may differ in the frequencies of some genetic variants, or they may not. If they do differ, they are racially distinct. Human races are neither more nor less objective or "real" than races in other species (Dobzhansky, 1970, p. 300).

Any scientific treatment of the question of human races stirs a host of emotional reactions. Race is a common, innocent term for use with other animals, but heavily loaded with prejudice and social connotations in humans – in particular after the Holocaust. Any use of the word race in respect to humans immediately brings to mind racial discrimination.

In the 19th century, and earlier, no such connotations disturbed the biologist's mind. Six principal races were recognized – dating from the early taxonomic attempts of Linneus: Negroid (African), Caucasian (European), Mongoloid (far-eastern, including Japanese, Chinese and Mongols),

Australian (Australian aborigines), Indian (American Indians) and Polynesian (Islanders of the Pacific Ocean). These divisions were based on morphological differences in skin color, shape of the head, and hair, and also on geographical distribution. In fact, some investigators subdivided humans into many more races – 30 are listed by Dobzhansky (1955). Darwin gives a list of authors who recognized up to 63 races in man. Some wanted to see this diversity of races as evidence for a polyphyletic origin of man, but modern investigators agree – with Darwin – that all human groups are descended from a single origin.

Darwin dedicated chapter VII in his 'Descent of Man' to the discussion of human races. He disagrees with those who consider the races distinct species: inter-racial marriages are fertile, and in some countries – notably Brazil – people of different races live together and freely intermarry. An important fact is, to Darwin, that the distinctive racial characters like skin color, hair, and facial structure, are all very variable <u>within</u> races, not only among modern men but also in primitive tribes. Darwin considers the alternative of calling these 'subspecies' rather than races, but decides against it:

> From long habit the term "race" will perhaps always be employed... Those naturalists who admit the principle of evolution – and this is now admitted by the majority of rising men – will feel no doubt that all races of man are descended from a single primitive stock (Darwin, 1874, p. 192).

Since the 1970s, investigators have agreed that the differences of races at the molecular level (in proteins or DNA) are very small, and that the observed morphological differences are determined by very few genes, which are not characteristic of the rest of the genome. Lewontin (1972) used data on blood-group polymorphisms and some electrophoretic markers to measure similarity in human populations. Similarity was measured by the Shannon information index ($H = -\Sigma \mathbf{p}_i \ln \mathbf{p}_i$, where $\mathbf{p}_i$ = the frequency of the allele i in the group). Lewontin's calculations showed that the differences among races amounted to just 6% of the total variance of marker frequencies. Population means within races contributed 8% more, but 84% of the variation was due to genetic variation among individuals within populations (Lewontin went as far as to suggest that the term race should be avoided, since it carries negative connotations). These results were supported by Nei (1978) and Nei and Roychoudhuri's (1974, 1982) electrophoretic analyses using Nei's genetic distance as a measure.

EUGENICS

One of the oldest and most controversial issues in human evolution is the protection of heritable deleterious characters by medical care. In the 19th century, Darwin's cousin Francis Galton was worried that medicine preserves the weak, and allows the sickly to survive. This 'compassion instinct' is a result of the human social sense, and illustrates the nicest aspects of human nature, but Galton – and many later 'eugenics' supporters – believed that it causes damage to the future of the human species. In primitive societies, the weak and the sickly were selected out: modern humans build hospitals to take care of them. "It is easy enough to see how quickly a race of domestic animals loses their good qualities when selection is relaxed!".

The anti-militarist biologist, Ernst Haeckel, saw the problem from a unique angle:

> In the past, selection was practiced in human populations (e.g., the ancient Spartans selected their newborn babies). Military selection [operates] in modern times, the strong and healthy young are recruited to be killed in wars, while the feeble remain and reproduce. Also medical help prolongs the life of the sickly parents, who transmit to their children many diseases (Haeckel, 1876 I: 170-175)

Galton left money in his will to establish a Chair of Eugenics at Cambridge University, which was given to the mathematician-philosopher Karl Pearson. The question of improving the genetic qualities of the human populations (positive eugenics), or at least protecting human populations from the accumulated effects of deleterious mutations, were the center of the Eugenics movement in the 20th century. All laws and customs favoring eugenics involved a certain amount of coersion and enforcement by law, when education and counselling failed. In the name of noble purposes, combined with racial prejudices and hatred, a large amount of injustice and cruelty were excercised in the USA (see Gould, 1981). In particular, the Eugenics ideals provided the philosophical foundation for the racial ideology in Nazi Germany in the 1930s and 1940s, leading to the atrocities against Jews and other minorities during the Second World War.

Whole symposia were dedicated to this dilemma in the mid-20th century. Arguments were raised that the negative effects of medical care may not be as serious as some people feared (Bajema, 1969). Carriers of genetic diseases which were formerly lethal, are kept alive and well by modern medical treatments. But what we conceive as ailments are the phenotypic effects on the patients, not the defective genes, and it is the phenotypic expression which is suppressed by medicine. The quantitative effect of medicine on genotype frequencies is not clear.

Some people argued that only those genetic diseases where the patient's life can be preserved with relatively minimal cost, like diabetes, may increase in the absence of selection, while genes that cause great harm, but for which the cost of treating the phenotype is heavy, will not be seriously protected by medical treatment.

There are serious objections to the entire eugenic approach: Who is to decide which genes or phenotypes are positive and should be protected, and which should be selected against? And, in principle, is it desirable to select a few 'good' genotypes (supposing that we know or can agree on what is a 'good' genotype) – or it is better to keep the human population variable?

> The genotype of a great benefactor of mankind, of a great national leader, of a great scientist, or of a saint, may result in a tyrant, a criminal, or a bum…
>
> Thus trying to multiply the Einsteins, the Lincolns and the Gandhis, we might obtain instead Stalins, Hitlers and Rasputins. (F. Ayala)

MAN AND THE ENVIRONMENT

There is no doubt that man has a great impact on the environment. Some changes are imposed by man to improve his standard of living: building of cities, bridges, highways, railroads and airports to meet the demand of the growing human population are examples. Other, harmful, changes are indirect but no less conspicuous: the dumping of waste in the rivers and oceans is a cause of great concern ("the conclusion, that dilution is the solution for pollution, is an illusion"). A corollary of this is the danger from pesticides and other poisons, dumped into the environment, reaching the water supply, accumulating in the food chain, and affecting non-target organisms, including man – and the list could easily be greatly extended.

The deleterious effects of man on his surrounding and the biological world were recognized (by some) long ago. Charles Lyell wrote in 1830:

> We can only estimate the revolution caused in the animal world by the growth of the human population – even only as consumer of organic matter (Lyell, 1853, p. 681)
>
> The draining of marshes, and the felling of extensive forests, caused such changes in the atmosphere as greatly to raise our conception of the more important influences of these forces. Before these changes the climate could be expected to be stable for thousands of years.
>
> The claims of some, that globally man's agricultural activities improve the productivity of the land, are wrong, because they are so much in the habit of regarding the sterility or productivity in relation to the wants of man, and not as regards the organic world generally (ibid., p. 681).

Not much could be added to these statements today, except that the situation is much worse than it was 175 years ago.

All animals and plants change the environment in some ways. Ants, rodents and plant roots dig tunnels and pathways as humans do. The activity of termites as builders of nests a few meters high, from soil and plant matter cemented by their saliva, are no less impressive than the building of sky-scrapers of which man is so proud. Ants that keep aphids as 'farm animals' which they 'milk' to get their sweet excreted honeydew, or *Atta* ants growing fungus gardens to feed on, are in principle very similar to human activities – except for their smaller quantitative effects on the global environment.

The difference between man and the animals is quantitative, not qualitative. The impact of a city and its millions of inhabitants is thousands of times greater than accomplished by the same number of insects or rodents. In his efforts to tie nature to his needs, man is destroying the balance – if this existed – between animals and the global resources, without building up a new balance in which man himself is a part.

THE EVOLUTIONARY FUTURE OF MAN

It must inevitably follow that the higher – the more intellectual and moral – must displace the lower and more degraded races... while his external form will probably ever remain unchanged, ...his mental constitution may continue to advance and improve, till the world is again inhabited by a single, nearly homogeneous race, no individual of which will be inferior to the noblest specimen of existing humanity (A.R. Wallace, 1870, p. 185).

These predictions, written 135 years ago by the idealist Wallace, sound detached from reality. The pessimistic opinions voiced in 1798 by Malthus, about the inevitable 'struggle for existence' – which formed a starting point for the theory of natural selection – seem to be closer to the truth. The geometric population growth predicted by Malthus in 1798 is certainly continuing:

The population problem has entered a new phase. It is no longer a race between population and food production, but between death-control and birth-control (Huxley, 1957, pp.178-9).

If nothing is done to control this flood of people, mankind will drown in its own increase... or the world economy will burst at the seams, and mankind will become a planetary cancer (ibid., p.181).

Human populations must be changing genetically with time. Some of these changes are subtle and usually go unnoticed. For example, radioactivity is used in energy production and in medicine, despite being mutagenic. Large quantities of radioactive waste are transported across the world and disposed of by unsatisfactory means. Radioactive wastes enter the air and the water supply (not to mention nuclear disasters like Chernobyl, nor the threat of the thousands of nuclear warheads in the possession of armies). Little is known of the cumulative effects of this radioactivity on human populations.

Genetic engineering is a growing industrial and research activity, utilizing the most recent advances in molecular technology. Organisms (particularly cultivated plants) are being modified for resistance to diseases, faster growth and higher yields, with some spectacular successes – amid warnings of the possible ecological dangers from the spread of genetically-engineered organisms.

The Human Genome project, recently completed, raised hopes for exact genetical location and the ultimate cure of genetic diseases in humans. The vehicles for such repair may be viruses, carrying the normal gene to the defective target site. Are the benefits greater than the risks? Although the possibility of human cloning has been seriously considered after the success with farm and laboratory animals, repair of genetic defects by genetic engineering has still a long way to go.

Literature Cited

Bajema, C.J. (ed.) 1969. Natural Selection in Human Populations. Wiley, New York, USA.

Darwin, C. 1874. The Descent of Man. 2nd ed. Hurst & Co., New York, USA.

Dobzhansky, T. 1955. Evolution, Genetics, and Man. Wiley, New York, USA.

Dobzhansky, T. 1961. Mankind Evolving. Yale University Press, Boston, USA.

Dobzhansky, T. 1970. Genetics of the Evolutionary Process. Columbia University Press, New York, USA.

Eckland, B.K. 1972. Evolutionary consequences of differential fertility and assortative mating in man. Evolutionary Biology 5: 293-310.

Galton, F. 1962 (1864). Hereditary Genius. The World Publishing Co., Cleveland, USA.

Goldschmidt, R. 1940. The Material Basis of Evolution. Princeton University Press, Philadelphia, USA.

Gould, S.J. 1988. Trends as changes in variance: a new slant on progress and directionality in evolution. Journal of Palaeontology 62: 319-329.

Gould, S.J. 1981. The Mismeasure of Man. Norton, New York, USA.

Graur, D. and W-H. Li. 2000. Fundamentals of Molecular Evolution. Sinauer, Sunderland, USA.

Haeckel, E. 1876. The History of Creation. Appleton, New York, USA.

Huxley, T.H. 1900 (1864). Man's Place in Nature and other Essays. MacMillan, New York, USA.

Huxley, J. 1957. New Bottles for New Wine. Harper, New York, USA.

Le Gros Clark, W.E. 1964. The Fossil Evidence for Human Evolution. 2nd ed. University of Chicago Press, Chicago, USA.

Lewontin, R.C. 1972. The apportionment of human diversity. Evolutionary Biology 6: 381-398.

Lyell, C. 1853. The Principles of Geology (9th ed.). Murray, London, UK.

Neel, J.V. and W.J. Schull. 1972. Differential fertility and human evolution. Evolutionary Biology 6: 363-379.

Nei, M. 1978. The theory of genetic distance and evolution of human races. Japanese Journal of Human Genetics 23: 341-369.

Nei, M. and A.K. Roychoudhuri. 1974. Genic variation within and between the three major races of man, Caucasoids, Negroids, and Mongoloids. American Journal of Human Genetics 26: 421-443.

Nei, M. and A.K. Roychoudhuri. 1982. Genetic relationships and evolution of human races. Evolutionary Biology 14: 1-60.

Waddington, C.H. 1957. The Strategy of the Genes. George Allen & Unwin, London, UK.

Wallace, A.R. 1870. Natural Selection. / Tropical Nature. MacMillan, London, UK.

Wallace, A.R. 1889. Darwinism. Macmillan, London, UK.

Weismann, A. 1891. Essays on Heredity. Oxford University Press, Oxford, UK.

Wilberforce, S. 1860. On "The Origin of Species by means of Natural Selection, or the Preservation of Favoured Races in the Struggle for Life". By Charles Darwin, M.A., F.R.S.London 1860. Quarterly Review 1860, 225-264.

Wool, D., and M. Burstein. 1992. Great tits exploit aphid galls as a source of food. Ornis 23: 107-109.

CHAPTER 25

Strategies in Evolution

Palaeontology teaches us that many species of animals were widespread and common during long periods of the Earth's history, but became extinct at some point in time. Their remains can be studied in museums of natural history. By contrast, there are a few species which have survived apparently unchanged, for very long periods indeed, such as the Brachiopods, the horseshoe crab (*Limulus*), *Latimeria* and lungfishes. The crocodiles changed very little from their ancient reptile progenitors. And of course all living species today are, by definition, successful survivors of evolutionary processes.

What, if anything, can we learn from a comparison of the characters of the surviving with the extinct species, about what may have enabled the survivors to succeed in evolution? To answer this question, we must first define what we mean by success.

THE EVOLUTIONARY GAME

One approach to this question is through the use of game theory. The course of evolution is described as a game between the population (species) of organisms and Nature. Nature presents the species with challenges (moves) in the form of changing environmental conditions. The species must respond to each move with a counter-move: for every environmental change, a suitable response by the species is required.

The sequence of steps taken by a player during the game is called its strategy. The strategy of Nature is the temporal sequence of environmental conditions. The evolutionary strategy of a species is its genetic composition: each genotype within a population has a range of possible responses or tolerances for a set of environments, and the response of the population – and the species as a whole - is the sum of the responses of its component genotypes.

THE MEANING OF SUCCESS

In any game, players play to win. In an ordinary game between humans, the winner gets some reward or prize: money, a cup, or prestige. A characteristic of these prizes is that it is impossible to use the prize to win another game. Prize money won in a game of billiards or golf can only be spent away from the game table or the field. The winning team in a basketball or a football championship match can put the cup on display to gain prestige, but cannot use it in the next match when it defends its title. (A famous story tells of a man who developed to perfection the special skill of throwing lentils through the eye of a needle, and when he demonstrated his skill before the king, he was rewarded with a bagful of lentils. This reward could be used on the arena – but obviously this was not the reward he wished to get.)

The evolutionary game differs from other games in one important aspect. In the evolutionary game, the player cannot get off the playing ground to enjoy its prize: "getting off the table" in evolution means extinction (Slobodkin, 1968). The game of species against nature is therefore similar to a card game in which the loser is immediately executed. Success means, remaining in the game.

The question of success in evolution can be rephrased: what strategy will give the population the highest probability of winning the evolutionary game – i.e., surviving for the longest time possible?

The strategy of Nature with which the species must cope is an unpredictable series of changes (nature does not reveal its cards). Lewontin (1961) examined three possible strategies for staying in the game. Each strategy is best in some environments but not in all. 1) Concentrate on a single genotype, best adapted to the set of conditions that occurs most of the time. This strategy carries the risk of losing the game when the conditions change. 2) Concentrate on a single, flexible genotype, which is able to adjust to a wide range of possible conditions. This genotype could be some exceptional homozygote which carries a strongly-linked assemblage of 'good' genes. (This model was suggested as an explanation for the widespread genetic (electrophoretic) monomorphism in the snail, *Rumina decollata*, which occupies very different habitats in the USA (Selander and Kaufman, 1973; Selander and Hudson, 1976).) 3) Construct a <u>variable</u> population, which will survive by changing its strategy (= genetic composition) in response to the environmental changes.

The choice of the best strategy depends on the strategy of Nature. In most of the theoretical models in the 1930s, Nature was assumed to present a constant strategy for a reasonably long time relative to the generation time of the organisms. In this case natural selection could adapt the population slowly to the new environment – in other words, increase its average fitness by a slow change of its strategy (allele frequencies). Even if the environment was not constant but changed cyclically, the organisms may still track the changes if the cycle takes longer to complete than generation time.

In the 1960-1970s, many biologists considered that constant environments were rare. Even in places which were formerly considered very stable ecologically – like the bottom of the ocean – some dimensions of the environment may still be variable (Ayala, 1975).

COARSE- AND FINE-GRAINED ENVIRONMENTS

Ecologists talk about 'fine-grained' and 'course-grained' environments (Table 25.1). In a coarse-grained environment, an organism may spend its entire life in the same environment, which it will conceive as constant (although the 'grains' experienced by different individuals may not be the same). An example may be a parasite, which spends its entire life in the body of a single host individual. A fine-grained environment is one where the individual must pass through several environments during its lifetime, and therefore will perceive its lifetime environment as variable. Examples

Table 25.1 Ecological and genetical strategies of organisms in coarse- and fine-grained environments

Description	ecological	genetical
Environment:	**constant**	**coarse-grained**
Expected response:	k-selection	genetic polymorphism
	Large size	
	Low reproductive output	
	Long development time	
	Long adult life span	
Environment:	**variable**	**fine-grained**
Expected response:	r-selection	single, flexible genotype
	Small size	
	High reproductive rate	
	Short developmental time	
	Short adult life span	

of fine-grained environments are provided by the invertebrates of the tidal zone or the metamorphosing insects and amphibians.

Ecologists labeled the organisms living in coarse- and fine-grained environments as k-strategists and r-strategists, respectively, and worked out the characteristics of each strategy (Table 25.1).

Geneticists offer their own suggestions for success in the two kinds of environments. In the coarse-grained environment, a strategy of a different genotype (or group of genotypes) in each grain will be advantageous – the population as a whole should be polymorphic. Conversely, in a fine-grained environment, a single, flexible, genotype should be favored (Levins, 1968).

Fine-grained environments are the more difficult for organisms to adapt to. These include sudden, unpredictable catastrophes (Slobodkin, 1968).

Lewontin (1961) constructed a simple model for comparing the different strategies that organisms may use to succeed in a fine-grained environment. The model assumed one gene with two alleles. Each allele has a constant value ('utility') in each of two alternative environments, wet and dry (Fig. 25.1). The heterozygote is assumed to be intermediate between the two in both environments.

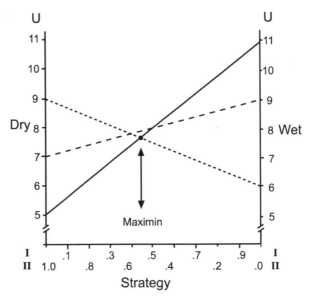

Fig. 25.1 Utility of different strategies of three genotypes in a randomly-alternating environment (after Lewontin, 1961). In this example, genotype AA (I, solid line) is wet-adapted ($U = 11$ in the wet, but only 5 in the dry environment). Genotype aa (II, dotted line) is dry-adapted ($U = 9$ in the dry, 6 in the wet). The heterozygotes Aa (broken line) are intermediate. The maximin strategy (see text) is indicated by the arrow.

The population is faced with wet and dry environments in random, unpredictable order. The probability of wet conditions is **p**, the probability of dry conditions is $1 - p$.

The utility to the population of a 'pure' strategy AA is calculated as follows:

$$U_{AA} = p^* U_{AA, wet} + (1 - p)^* U_{AA, dry}$$

A similar formulation is used for pure strategies of aa and Aa. The utility of a mixed strategy, including more than one genotype, can be calculated similarly (Lewontin, 1961) (Fig. 25.1).

These calculations showed that a mixed strategy (a polymorphic population), is a 'maximin' strategy: it has the lowest utility when conditions are maximally favorable, but the highest average utility when conditions are bad. When the purpose of the game is to stay in it for the longest possible time, the maximin strategy is the more desirable: so long as the population survives when conditions are the worst, it does not matter how well it does when conditions are better!

In Lewontin's simulations, mixed combinations of the homozygotes AA and aa (with no heterozygotes) gave the highest maximin utility. He suggested that when the environment changes at random, the best strategy is a geographically-variable population, with each cell homozygous for a different allele. To ensure homozygosity, each cell should be highly inbred or even selfed.

CONCLUDING REMARKS

As I interpret the theory [of evolution], it is truly scientific, for the very reason that it does not attempt to explain anything. It takes the facts of life as we perceive them, and attempts to describe them in a brief formula, involving such concepts as "variation", "inheritance", "natural selection" and "sexual selection" (Karl Pearson, 1900).

Natural selection has no analogy with any aspect of human behavior. However, if one wanted to play with a comparison, one would have to say that natural selection does not work like an engineer works. It works like a tinkerer – a tinkerer who does not know exactly what he is going to produce, but uses whatever he finds around him, whether it be pieces of string, fragments of wood, or old cardboard. In short, it works like a tinker who uses everything at his disposal to produce some kind of a workable object (Jacob, 1977).

The analogy of the processes of natural selection with the operation of a tinkerer is an accurate description of the most important driving force in evolution. The combination of chance effects (genetic drift) with the operation of natural selection thus described, accounts for the observed fact that organisms are operational units and their parts work together. No planning is required to produce the observed result. Evolution is not entirely governed by chance processes, because natural selection builds on structures already present, and constructs new forms from old materials.

T.H. Huxley wrote in 1880, that the new generation '20 years from now' may accept the doctrine of evolution by natural selection uncritically "perhaps, in the same unjustified consensus as our (Huxley's) generation rejected it". It is interesting to speculate to what degree this prognosis is true now, 125 years later and 145 years after the publication of the Origin.

It seems that a considerable part of the world's population who have heard of Darwin and his theory (excluding perhaps the greatest part of humanity who have not), still object to the theory of evolution for the same reasons that were raised by Samuel Wilberforce in 1860. The theory however is still valid as it was then. The progress of molecular technology and the prospects of genetic engineering for solving all the genetic maladies should not obscure the fact that molecular evolution can only take place via changes in the proportions of individuals carrying different genotypes – regardless of whether the mutated genes have differential effects on the fitness of the organisms, or if these genotypes were fixed by chance.

Basically, evolution is still a slow process, which affects individual organisms. I believe that we should return to the study of whole organisms in the natural environment – the ecological-genetic paradigm. An organism is not a collection of genes: it is a unit which interacts with the environment in various ways. DNA sequences are very useful as markers for individuals, but the responses of individuals cannot be predicted from their DNA sequences.

I stated earlier that the effect of man on the environment is not qualitatively different from the effect of other organisms. But the quantitative effect is so great that it threatens to destroy the biological world, man included. Human population growth, combined with the destruction of the natural environment and the resulting pollution of what remains, must get the world leaders to think of the future.

The technological advances should be used to restrain the destructive processes. The same cultural evolution that created the most noble concepts of culture – morality, love of mankind, and sanctity of human life – should be strong enough to ensure that the destructive processes will not annihilate the biological world.

Literature Cited

Ayala, F.J. 1975. Deep-sea asteroids: Hugh genetic variability in a stable environment. Evolution 29: 203-212.

Jacob, F. 1977. Evolution and tinkering. Science 196: 1161-1166.

Levins, R. 1968. Evolution in Changing Environments. Princeton University Press, Philadelphia, USA.

Lewontin, R.C. 1961. Evolution and the theory of games. Journal of Theoretical Biology 1: 382-403.

Pearson, K. 1900. The Grammar of Science. 2nd ed. A. & C. Black, London, UK.

Selander, R.K., and D.W. Kaufman. 1973. Self-fertilization and population structure in a colonizing land snail. Proceedings of the National Academy of Sciences, USA, 70: 1186-1190.

Selander, R.K., and R.O. Hudson. 1976. Animal population structure under close inbreeding: the land snail *Rumina* in southern France. American Naturalist 110: 695-718.

Slobodkin, L. 1968. Toward a predictive theory of evolution. pp. 187-205. In: R.C. Lewontin (ed.), Population Biology and Evolution, Syracuse University Press, Syracuse, USA.

Author Index

Index

Printed and bound by CPI Group (UK) Ltd, Croydon, CR0 4YY

28/10/2024

01780249-0003